GLOSSARY OF *Typesetting* TERMS

Glossary of *Typesetting* Terms

· ·

Richard Eckersley

Richard Angstadt

Charles M. Ellertson

Richard Hendel

Naomi B. Pascal

Anita Walker Scott

The University of Chicago Press *Chicago & London*

The University of Chicago Press,
Chicago 60637
The University of Chicago Press, Ltd.,
London
© 1994 by The University of Chicago

Published 1994
Printed in the United States of America
03 02 01 00 99 98 97 96 95 94
1 2 3 4 5
ISBN 0-226-18371-8 (cloth)

⊛
The paper used in
this publication meets
the minimum requirements
of the American National
Standard for Information
Sciences – Permanence of
Paper for Printed Library
Materials, ANSI Z39.48-1984.

Library of Congress Cataloging-in-Publication Data
Glossary of typesetting terms / Richard Eckersley . . . [et al.].

    p.   cm. – (Chicago guides to writing, editing and publishing)
    Includes bibliographical references.

    1. Type-setting – Dictionaries.   I. Eckersley, Richard.
II. Series.
Z253.G57   1994
686.2'2503 – dc20   94-21223
                                       CIP

CONTENTS

The development by Johann Gutenberg, in the mid-fifteenth century, of movable type, printing presses that could sustain heavy work, and the various processes and manual operations of printing was among the most significant contributions in world history. In the long span of time since then, printed text has become a powerful communication tool. In their shapes as well as their content, words affect every aspect of our lives. And words are everywhere, all around us, in countless forms, sizes, and combinations.

Gutenberg's ingenious resolution of complex mechanical problems into a practical method of typesetting and printing established both the process and the language of type. For the next four hundred years that language hardly changed. Late in the nineteenth century, the Linotype and Monotype typesetting machines were introduced. Although they set type at great speed, they still relied heavily on Gutenberg's vision and the principle of lead type, and much of the vocabulary relating to materials and application remained unchanged. Words that were unique to the new mechanical systems were simply assimilated into the existing type vocabulary. In the late 1940s, however, with the introduction of a composition system that used photographic means, the language of typesetting, which until this time had been clearly understood and generally accepted, was suddenly and dramatically altered.

Along with new terms, a whole new way of thinking about what was typographically possible had to be learned. The accepted arcane and jargonistic vocabulary of metal typesetting had to be either cast aside, adapted, or redefined to fit the changes in technology. When computer-oriented composition arrived with force and flair in the 1980s, those who worked with type adapted to change once again. And this time the differences in approach, attitude, and language were profound.

At each stage of development in typesetting machinery, the best manufacturers made an effort to preserve the basic terminology that had originally been developed for metal typesetting. Over time, however, each new system put a different accent on the language, sometimes subtle, sometimes striking. Many meanings, rather than being universal, have become parochial, with narrow, specialized definitions assigned to some terms while broader meanings are accepted for others. Often, shared understanding of the type vocabulary depends on the users' age or experience.

Moreover, digital technology has made the transformation of words into type accessible to anyone with a computer system and the proper software. Clear understanding of type language is not really essential. No long apprenticeship, no previous experience with type or design, not even a basic understanding of the tradition and history of type is required. Only typing skills are needed to give access to almost unlimited type choices and arrangements.

For all practitioners of an art, regardless of the medium, the value of their technical vocabulary depends both on precision in defining terms and on the assumption that everyone who uses the words shares the same understanding of their meaning. For too many years now the precision has been lacking and the assumption has been false. Yet the language and craft of typesetting are securely grounded in typographical standards and traditionally accepted principles. These standards can be applied equally to any method of typographic composition, and their importance cannot be overestimated. Understanding this foundation and the related terminology can make the difference between mediocre and distinguished typographic work.

The task force responsible for this book had its beginnings in 1990 at the annual meeting of the Association of American University Presses in Philadephia. We were members of a panel discussing typesetting and the difficulty of writing and translating specifications so they could be universally understood. We also offered possible solutions. At the conclusion of our presentation, it was clear that a need had been identified. Audience awareness was keen, and the questions asked were perceptive and intriguing. Everyone seemed enthusiastic about creating a typesetting glossary. Before we left Philadelphia, the task force had been formed and given its official charge by the AAUP board of directors.

Our goal in preparing this glossary has been to define the language of typesetting in terms of the present technology while also providing, where appropriate, applicable standards and historical context. The volume is intended for designers, typesetters, editors, proofreaders, and related professionals as a reference and guide to better communication and as an interesting journey through the eloquent language of type.

During the four years of preparation, our discussions inevitably reflected the points of view and experience of the task force members – three designers, an editor, and two typesetters. We attempted to forge a common language from our distinct professional dialects and the personal accents within each of them. We did not try to be exhaustive – to trace the technical meaning of every term through all the widely used composition systems, possibly ending up with a "consensus definition." We by no means always agreed on the "best" definition, but we did try to determine meanings as they apply in current usage. At times, as strong and perhaps stubborn professionals, we agreed to disagree, acknowledging that we have specific preferences but accepting majority opinion for the definition of individual terms. In some cases we have attached personal comments, identified by the author's initials.

You will find differences between this glossary and the terminology in the respected *Chicago Manual of Style*. In a field so rich, diversity of opinion is inevitable. In a technological graphic arts world that is changing even as this book goes to press, fallibility is certain.

The original list of terms was much longer than the one that appears here. Typesetting terms that were obscure, archaic, self-explanatory, or redundant were eliminated. In these categories were many words related to earlier typesetting practices that no longer are current. Though we made many adjustments in focus, the goal has remained the same – to improve communication between the originator and the interpreter of type specifications.

During the years of preparation, we have benefited from the cooperation and help of many colleagues. Carl Gross of the University of Pennsylvania Press and Steve Renick of the University of California Press contributed to initial list making and planning. Veronica Seyd and Julidta Tarver of the University of Washington Press, Larry Tseng of Tseng Information Systems, and Linda Angstadt of Keystone Typesetting provided valuable assistance at various stages of the manuscript. Bill Grosskopf and Richard Workman of G & S Typesetters allowed us to reproduce some of the house style forms they have developed and gave generously of their experience and expertise. Peter C. Grenquist, executive director of the Association of American University Presses, was always supportive and encouraging and was instrumental in arranging for publication. Acting as readers and critics of the complete manuscript were two designers and typographic experts. Dwight Agner has worked intimately with every system of composition from handset to desktop, and he applied that knowledge to his reading. George Mackie, for many years associated with the University of Edinburgh Press, contributed his talents and experience in commenting on the manuscript. At the University of Chicago Press, John Grossman, then managing editor, generously shared with us portions of the manuscript of the fourteenth edition of *The Chicago Manual of Style;* Penelope Kaiserlian, associate director, assisted us through the final preparation of the work and granted permission to reproduce the table of proofreaders' marks from the *Chicago Manual;* Alice M. Bennett, senior manuscript editor, provided careful copyediting of the final draft; and Sylvia Hecimovich, assistant production manager, arranged and supervised production.

My final thanks go to the other members of the task force. They have been unstinting with their time, knowledge, and enthusiasm in spite of full-time dedication to each of their responsible positions.

Our journey together has been one of intellectual exchange, learning, and friendship.

The *Glossary of Typesetting Terms* brings together our varied experience, our passion for type and design, and the responsibility we feel toward the education of those who use and appreciate the language of type.

Anita Walker Scott
Chair, AAUP Production Glossary Task Force

Glossary

NOTE
Words in **boldface** are cross-references to separate entries.

An italic comment following an entry was written by the individual author whose initials appear at the end.

AA. Author's alteration. *See also* **Alteration.**

AC. Author's correction. *See also* **Alteration.**

Accent. Mark over, under, or through a letter that modifies its sound, stress, or pitch, in accordance with the practices of the language; often used interchangeably with **diacritic.** *See also* **Appendix 7.**

Acknowledgments. Part of the **front matter** in which the author expresses appreciation for assistance in research, in writing the manuscript, in producing the book, etc. Acknowledgments may appear either as a separate section following the **preface** or – if they are not too long – at the end of the preface itself. Acknowledgment of permission to use previously published or copyrighted material may be included there or may be placed on the **copyright page.** *See also* **Appendix 2.**

Acute accent (´). *See* **Accent; Appendix 7.**

Ad card. A list of previous books by an author, usually placed in the **front matter.** *See also* **Appendix 2.**

Addendum (plural *addenda*). Material added to a publication at too late a stage in production to be incorporated into the text; usually inserted as a loose sheet, but sometimes attached by binding or gluing. *See also* **Corrigendum; Erratum.**

Aesc. *See* **Ash.**

Agate. (1) A type size of approximately 5½ points, 14 lines to the inch. Now used most commonly as a unit of measurement in newspaper advertising. (2) *Agate line,* a unit of measurement equaling ¹⁄₁₄ of one **column inch.** *See also* **Body, body size.**

A-head. The primary order of **subhead** in a text. *See also* **Appendix 4.**

ʿ**Ain.** *See* ʿ**Ayn.**

Album format. *See* **Oblong.**

Align as typed. Instruction particularly common in marking verse, where the intention is to preserve in the typesetting the visual pattern of the manuscript.

The result of this instruction may not be what was intended, because of the different spacing characteristics of typewritten and typeset copy. A monospace typewriter, with every keystroke allocated the same width, creates a natural vertical alignment of letters and spaces. Since the typesetting machine allocates varying spaces to letters and letter pairs, with comparatively tighter word spacing, the vertical relationships of letters and the relative lengths of lines may be very different from those in the manuscript. Thus, in a stanza pattern where the manuscript has all but the first and last lines indented one character space, the indention should be specified in typesetting terms of ems and half ems; the instruction "align as typed" may produce too small a space. With complex line patterns, the copyeditor should identify lines of a common level of indention, for example, "align with line A above." Lines with a common indention may also be identified by drawing a vertical line. RE/CME

Aligning figure. *See* **Lining figure.**

Alignment. The vertical or horizontal relationship of elements (type, rules, figures, etc.) on the page. *See also* **Base align.**

Align on. Instruction to align one element in relation to another whose position has already been defined. For example, **chapter heads** may be aligned with reference to a particular line of the **text page,** as in "align chapter head on text line 5"; turnover lines may be indented to align vertically with a specified element above.

Align type. Instruction to correct the relative position of type vertically or horizontally.

Alphabet length. The length in points of the lowercase alphabet of a particular size and style of type. This information is supplied by type manufacturers for use in **copyfitting.**

Alphanumeric. Contraction of *alphabetic* and *numeric.* (1) Referring to the fifty-two characters of the upper- and lowercase alphabets plus the ten basic number characters 0 through 9. (2) Used in the context of electronic manuscripts to refer to the text portion as opposed to the code portion.

Alteration. A change marked on a **proof** that is a departure from the **manuscript,** as distinct from a correction made to eliminate an error by the typesetter. *Author's alterations,* sometimes called *author's corrections,* are identified on the proof by the code AA or AC and are usually charged – at least in part – against royalties. An alteration made by an editor is sometimes marked EA (*editor's alteration*). Similarly, a revision of typographic or other visual elements made in proof by the designer is often marked DA (*designer's alteration*). A change made by an editor or designer, not charged to the author, can also be marked HA (*house alteration*). A typesetter's mistake is called a PE (**printer's error**) or **typo** and is corrected by the typesetter without charge.

In hot-metal composition, the charge for alterations was usually by the affected line, since all such lines had to be rekeyed, cast, and proofread. Adding a single word to the first line of a paragraph might mean that the entire paragraph would have to be reset, and the alterations charges would count every line in that paragraph. Although the practice of charging by the line has to some extent been carried over into photocomposition, there is a trend toward charging by "occurrence," since most photocomposition systems will permit the typesetter to reset the individual change without having to rekey and proof the whole paragraph. CME

Alternate characters. Different versions of the same character, usually in **display faces** or sizes; for example, **swash letter.**

Ampersand (&). Character derived in form from the Latin *et,* meaning "and." In **Old English,** the character ⁊ or 7.

Angle brackets. *See* **Brackets.**

Anglo-Saxon. *See* **Old English** (2).

Antique. Style of type with heavy vertical strokes of optically uniform thickness and square serifs, exemplified by the typefaces Figgins Antique, Thorne, and Clarendon. Used widely in the nineteenth century for setting playbills, and still popular for display setting.

Apostrophe ('). Mark of punctuation used to denote the possessive or to indicate the omission of a letter or letters. The same character is used for a single closing quotation mark.

Appendix. Section of the **back matter** of a book containing material that is not part of the main text but provides useful background or further clarification. *See also* **Appendix 2.**

Arabic numeral. Any one of the digits 0 through 9, as distinct from a **roman numeral.** *See also* **Lining figure; Old-style figure; Pagination.**

Artwork (A/W). Any material ready for reproduction. In bookwork, any element the typesetter is not required to furnish (such as an exotic initial or an antique ornament) is identified in the type specifications as "A/W to be supplied" (that is, supplied by the publisher). The term *artwork* is also used to distinguish any element that cannot be generated totally on the typesetting machine but requires additional handwork (inking, cutting, or pasting).

Ascender. The upper portion of a lowercase letter that extends above the **x-height,** as in b, d, f, h, l. *See also* **Appendix 1.**

ASCII. Acronym for American Standard Code for Information Interchange, pronounced "askey." An eight-bit computer code permitting 256 unique combinations. Each code signifies a particular **character.** The character set includes upper- and lowercase alphabet, numbers, punctuation, an assortment of common symbols and accented letters, and control codes. Used by most programs that run on personal computers.

ASCII file. Strictly, a computer file composed entirely of **ASCII** characters. ASCII covers a total of 128 characters. Of these, 95 are used in text and are sometimes referred to as "printable" characters. The remaining 33 "nonprinting" characters are commands that aid the transmission and reception of information between communicating devices. The existence of the standard makes files of this type very portable between different types of computers, programs, and operating environments.

ASCII files are sometimes referred to as "text only" files because the code does not directly support graphics and formatting beyond what can be accomplished with horizontal and vertical tabs, carriage returns, line feeds, form feeds, and backspacing. Programs that translate graphics and formatted text from their native form into ASCII generally ignore codes for which there is no ASCII equivalent; thus formatting and font changes may be lost. But ASCII files are not always text files, since the meaning of any character sequence ultimately depends on the recipient. An ASCII-encoded PostScript file, for example, uses only ASCII characters in its set of instructions for the PostScript printer. RTF (Rich Text Format) forms graphic commands using "words" that begin with an ASCII backslash – an arrangement readily understood by programs that recognize RTF. In both instances, the manufacturer selected the ASCII character set for its portability. ASCII-encoded PostScript files and RTF files are also ASCII files. CME

Ash, aesc (Æ, æ). Old English ligature, also used in Danish and Norwegian.

Asterisk (*). Typographic symbol used to indicate an unnumbered footnote. A double asterisk (**), sometimes used to indicate the second footnote on the same page, creates unsightly gaps in the word spacing of the line and should be avoided if possible. Single or, more commonly, triple asterisks are used to indicate a space break within a chapter. *See also* **Reference mark.**

In statistical matter, particularly tables, the number of asterisks has come to signify degree of significance. In such occurrences the use of one, two, and three asterisks should be preserved. CME

Author's alteration (AA). *See* **Alteration.**

Author's correction (AC). *See* **Alteration.**

A/W. *See* **Artwork.**

ʿAyin. *See* ʿ**Ayn.**

ʿAyn (ʿ), also *ʿain* or *ʿayin.* A **diacritic** used in setting transliterated Arabic and Hebrew. May be represented by a rough **breathing** from a Greek font or a single opening quotation mark.

The use of the opening quotation mark is acceptable only if it is from a font where the mark curves. Thus a Trump or Palatino opening quotation mark (') is not an acceptable rendition of the ʿayn, even if the text font is Trump or Palatino. CME

B · · · · · · · · · · · · · · · · · ·

Backing up. (1) Printing the reverse side of a sheet that has already been printed on one side; also called *perfecting*. (2) The relationship of the type areas on the front and back of a page when these are in register, exactly back to back.

Traditionalists consider backing up (2) a requirement of good book production, since it diminishes show-through or ghosting – type from the reverse page being visible through the paper. This is probably the reason such odium attaches to the practice of carding, since it produces situations where the line spacing differs on the two sides of the leaf. A distinction should be made, however, between letterpress and offset printing in this regard. In letterpress printing the show-through may be marked if the pages are not backed up, since the type is impressed into the paper. With offset printing, in which the ink is not impressed but lies on the surface, show-through is relatively slight. Consequently the subtle use of carding may be less than a criminal offense, or even desirable. RE

Back margin. *See* **Gutter margin; Margin.**

Back matter. The elements of the book that follow the main text and that may include a glossary, appendixes, endnotes, bibliography, index, colophon, or similar copy. Back matter usually begins on a **recto.** Also called *end matter* or *reference matter. See also* **Appendix 2.**

Backslash (\). A diagonal line or oblique stroke sloping downward from left to right. *See also* **Slash.**

Bad break. (1) A break in the setting of lines that offends against good editorial sense or aesthetics. In **page makeup,** starting a page with a paragraph ending of only a few words (a **widow**); ending a paragraph with a short or hyphenated word (an **orphan**); or ending a page with a **subhead,** a single line of an **extract,** or an unsignaled space break. (2) Incorrect hyphenation of a word at the end of a line.

Though now generally tolerated, the practice of ending a page with an indented paragraph opening weakens the rectangle of the type area and should be avoided if possible. RE

In computer composition, hyphenation is usually performed automatically via a single stored dictionary. If the publisher prefers "stand-ard" (Merriam-Webster, Second International*) to "stan-dard" (Merriam-Webster,* Third International *and* Tenth Collegiate*), the typesetter may fairly count the change as an EA, not a PE. Setting type is inherently a matter of making compromises; if the compositor has made reasonable ones, following the "general principles" set forth in* The Chicago Manual of Style, *the compositor should not be charged with a printer's error simply because the publisher has different preferences – unless, of course, these have been agreed on in advance. Other examples of bad breaks include dividing "p. 7" between the period and numeral or "G. E. Moore" between the initials.* CME

Bad copy. *See* **Penalty copy.**

Barred ell. *See* **Slashed ell.**

Barred oh. *See* **Slashed oh.**

Base align. To arrange type and other elements on a common **baseline,** usually that of the **text.** For example, footnote numbers at the bottom of a page may be specified as base-aligning figures (that is, sharing the baseline of the note text) rather than as superior figures. Also used when correcting proof, to indicate letters that have strayed out of alignment through some technical error. *See also* **Appendix 8.**

Baseline. The horizontal on which the letters of the alphabet optically range and below which the **descenders** fall. *See also* **Appendix 1.**

Baseline skip. *See* Leading.

Base to base (b/b, B/B, BB, or B to B). The vertical measurement of space between elements, from the **baseline** of one to the baseline of another. For example, the position of a subtitle in relation to the title above it may be specified as 32 **points** base to base. Generally, this is the method preferred by typesetters.

Bastard title. Usually the first page of a book, unless there is a **series title.** Sometimes called *false title* or *first half title.* The bastard title usually includes only the main title of the book, omitting the subtitle, author's name, and publisher's **imprint.** *See also* **Half title; Appendix 2.**

b/b, B/B, BB, or B to B. Abbreviation for **base to base.** In proofreading, B/B is sometimes also used for **bad break.**

Beard. Nonprinting area below the letter on the **body** of type.

Begin arabic. Instruction identifying the first page that should be numbered with **arabic numerals.** The previous pages (the **front matter**) are usually numbered with **roman numerals.** Traditionally, the **half title** is regarded as arabic 1, followed by a blank **verso,** and the first page of the main text is numbered 3. Some publishers prefer to begin arabic numbering on the **recto** page that follows the half title (the first page of the main text) in order to keep the option of dispensing with the half title later in the production schedule. This allows pages to be saved without renumbering the main text pages. *See also* **Appendix 2.**

BF. *See* **Bold, boldface.**

B-head. The order of **subhead** second in importance after **A-head.** *See also* **Appendix 4.**

Bibliography. List of sources cited or consulted, usually part of the **back matter** of a book. The bibliography may be in essay form rather than an alphabetical list, or there may be separate lists of references following each chapter of a multiauthor book. *See also* **Appendix 2.**

Biblio style. *See* **Flush and hang.**

Bitmap, bitmapping. With reference to a **font** or other image, bitmaps are always *rasterized* data, meaning they have been converted to a fixed quantity of information. (Thus a *bitmapped font* – often called a *screen font* – has letters constructed of square dots, each representing a dot on a computer screen.) Since rasterizing optimizes the rendering of an image for both a particular size and a particular **resolution** (of a display monitor or output device), the quality of a bitmapped image will be degraded if it is scaled (changed in size). In contrast, a PostScript file (also known as an outline file) contains mathematical instructions for rendering lines, curves, and angles. The images stored in a PostScript file are thus "elastic" and will maintain their integrity when output at any enlargement or resolution. In passing, when a PostScript file is sent to an output device or displayed on a monitor, it will perforce be rasterized – in effect, a bitmapped image will be generated at a particular size for a particular device. The original PostScript file, however, will remain "elastic."

Black letter. A group of typefaces derived from twelfth-century Gothic **minuscule** script. Also called **gothic** or **Old English** and used most often for period effect. Cloister Black is probably the best-known version readily available, Hermann Zapf's Textura one of the best drawn. *See also* **Appendix 3.**

Blank. An unprinted page. Term used in type specifications to describe a page without type or any other element.

The usual convention, when composition is billed by the page, is that any valid page (typically a page must have at least five lines of type) is billed at the page rate. Blank pages are not counted. CME

Bleed, bleeding. Term used to describe an image or a rule that extends beyond the indicated **trim** of the page. Thus a full-page illustration that has no margin of white would bleed on three sides (top, outside, bottom), not four – its continuation into the **gutter** is not a bleed, since the gutter side of the page is not trimmed. Printers prefer that any element that is to bleed, including typeset matter such as a **rule,** should continue ⅛ inch beyond the trim.

Blind folio. A page number that is counted but does not appear on the page.

Block. (1) Term used to refer to almost any definable typographical element on a page when giving instructions for inserting or moving, such as "block **FR**" or "block center" (*see also* **Copy block**). (2) *Block style* and *block paragraphs* refer to unindented paragraphs that are usually defined by extra space above and below.

Block quotation. *See* **Extract.**

Blueline, blue. As used in typesetting, a proof of the negative of a **halftone** or other graphic art, sometimes used as a *position print.* Although the method of producing the proof is different, the term is also sometimes used for *brownline, cyanotype, silverprint,* or *vandyke.* Also called *loose blue. See also* **FPO.**

Blurb. *See* **Jacket flap copy.**

Boards. In **pasteup,** the corrected **galley** (**repro**), cut and pasted onto card or heavy paper as single pages or double-page **spreads,** with all elements in position and ready for the printer. *Grid sheets* are boards with guidelines to aid the pasteup artist. These guidelines, preprinted in a blue ink (called *nonrepro blue*), will not appear when the boards are photographed, since lithographic film is not sensitive to nonrepro blue. *See also* **Mechanical.**

Body, body size. Originally, the piece of metal type that carries the letter, including the space below it called the **beard.** Thus, for 10-point type it is the body that measures 10 points; the letter size, measured from top of ascender to bottom of descender, is somewhat smaller, depending on the typeface. In most systems of machine setting (both metal and photographic), the size of the body includes the space added below the letter (**leading**) and is usually measured in points and half points. For example, 9½-point type with 3 points of leading (referred to as 9½ on 13) would be on a 13-point body, that is, having a body size of 13 points. *See also* **Appendix 1.**

Body, body type. (1) **Text** matter, as opposed to **display.** The term *body copy* has essentially the same meaning but is used more commonly in publicity and magazine work. (2) In **tables,** the columns of data as distinct from the identifying elements – **spanner heads** and **column heads.** *See also* **Appendix 6.**

Body copy. *See* **Body, body type.**

Bold, boldface (BF). A heavier weight of a standard typeface. For example, the words being defined in this glossary are set in boldface type. *See also* **Appendix 8.**

Border. A linear **ornament** in the form of one or more **rules** or a combination of separate ornaments. May be used vertically, horizontally, or to form the perimeter of a **box.**

Bottom margin. The space from the bottom of the type page to the trimmed edge of a page. Also called *tail* or *foot* margin. *See also* **Margin.**

Bottom of text, last line of text. The position of the last line of text on a page of regular depth. The vertical position of **folios** and **running feet** is specified in terms of their distance from the bottom text line, measured in **points** from **baseline** to baseline. *See also* **Bottom of type page; Text page.**

Bottom of type page. The position fixed by lowest element on the page – a line of text, a **foot folio,** or a **running foot;** not to be confused with **bottom of text.**

Bounce. Instruction to allow the page numbers at the bottom of the page to vary in position. *See also* **Bouncing folio.**

Bouncing folio. A page number that changes its position at the bottom of the page when the page is **short** or **long,** since its placement is in relation to the last line of text (except on pages that fall well short, as a chapter ending may). *See also* **Drop folio; Foot folio.**
Some designers hold that folios should bounce on long pages only, retaining their standard alignment on short pages. When this style is preferred it should be explicitly stated, since it is less common. CME

Boustrophedonic. Referring to an ancient way of writing that alternates in direction from left to right, then right to left (from the Greek term for the way an ox turns in plowing). For example:

As if to measure out his confusion by the mile, Ernest Bunbury paced the length of the library

from Abelard to Zola
bɹoɭǝq∀ oʇ ɒɭoⱬ moɹℲ
from Abelard to Zola.

Bowl. The rounded stroke of a letter, enclosing a **counter.** *See also* Appendix 1.

Box, box rule. An outline rectangle usually enclosing typeset matter or an illustration. Boxes can easily be generated on modern phototypesetting machines or computers. The thickness of the rule should be specified in points and half points, the width and depth of the box in picas and points.

Boxheads. Archaic term for headings in tables. Its use is discouraged, since it tends to conflate **column heads** and **spanner heads,** which may require different type styles. *See also* **Appendix 6.**

Box indent. The space formed for a **drop cap** by indenting the first lines of a paragraph the same amount. Also called *block indent.*

Brace ({ or }). A mark enclosing several listed elements to show that they are to be considered as a unit;

for example,
$$\left.\begin{array}{l} \text{rats} \\ \text{spiders} \\ \text{bats} \\ \text{wasps} \\ \text{toads} \end{array}\right\} \text{not nice}$$

Also called a *bracket. See also* **Brackets.**

Brackets. (1) Marks used in pairs to set off part of the text. *Square brackets* [ ] are sometimes used to enclose editorial interpolations or as a secondary enclosure within parentheses. *Curly brackets* { } may be used as a third level of enclosure. *Angle brackets* ⟨ ⟩ and square brackets can have special meanings in edited texts. In mathematical setting brackets are sometimes referred to as *fences;* the accepted sequence is ⟨ { [( )] } ⟩, although usage may vary. (2) The connecting curve between the **stem** of a letter and the **serif** (*see also* **Appendix 1**).

Break. (1) The place in the text where a line of type or a page ends. (2) In proofreading, an instruction to end a typeset line at a given point. *See also* **Bad break; Break block; Appendix 8.**

Break block. An instruction marked on proof to change line breaks in order to avoid the repetition of a word at the ends of consecutive lines or a run of consecutive hyphens or dashes. The objection to such word blocks is both editorial, in that they may cause readers to skip a line (and thereby lose their place), and aesthetic, in that they create **rivers** of white space. If *the* begins or ends three lines in a row, but one occurrence is in italic or begins with a capital letter, the block is effectively broken. Nevertheless, some publishers consider this situation a word block and require that it be broken. *See also* **Consecutive hyphens.**

Breathing (Latin *spiritus*). In Greek, a mark placed over a vowel that begins a word (before it when the vowel is uppercase), or over the second vowel of an initial diphthong. A rough breathing mark or *spiritus asper* (ʽ) indicates that an aspirated h sound precedes the vowel or diphthong: ἕν (*hen*), οἱ (*hoi*), Ἥρα (Hera). A smooth breathing mark or *spiritus lenis* (ʼ) indicates the absence of aspiration: ἐς (*es*), οὐκ (*ouk*), Ἀθηνᾶ (Athena). If a syllable with a breathing mark also has an acute or grave accent, the accent follows the breathing mark: ἥ (*he*). A circumflex is placed above the breathing mark: εἷς (*heis*). The rough breathing mark is also used over an initial rho (transliterated *rh*): ῥόδον (*rhodon*). A medial double rho is sometimes written with a smooth breathing over the first rho and a rough breathing over the second (transliterated *rrh*): Πύῤῥός (Pyrrhos). Breathing marks and accents are omitted from Greek words written entirely in uppercase. Vowels with every combination of breathing marks and accents are an integral part of any Greek font.

Breve. A mark used to show a short vowel or syllable. As a **diacritic** (˘), it is usually rendered smaller than as a **scansion mark** (˘). *See also* **Appendix 7.**

Broadside. Turned ninety degrees on the page; used for tables, maps, figures, plates, or illustrations too wide to fit horizontally on the page. Conventionally, the top of a turned element should be toward the left margin, to be legible when the book is turned clockwise. Also called *turned* or **landscape.**

Broken type. Proofmark indicating damaged type. A term inherited from metal setting, when type might have been imperfectly cast or have broken under the pressure of printing. Also called *cut type. See also* **Appendix 8.**

Brownline. *See* **Blueline, blue.**

Brush letter, brush script. (1) A letter drawn with a brush. (2) A style of type imitating brush-drawn letterforms (for example, Ashley Script).

Bullet (•). A heavy, solid dot usually centered vertically on the **cap height.** Used as an **ornament** or a point of emphasis, particularly in listed matter. To be distinguished from **centered dot.**

C ·

Callout. (1) Descriptive text or **label** included within the body of a diagram or figure to identify or explain its elements. (2) Editorial marking identifying footnote references in the text (*see also* **Footnote call**). (3) Editorial marking instructing the compositor where to place artwork.

Camera-ready copy (CRC). Any element ready to be photographed by the platemaker or printer. Also called *camera copy. See also* **Mechanical; Repro.**

Canceled type. Letters that are scored or marked with a cancel mark to indicate something crossed out in a document.

Cap height. The actual height of the capital letters in a **font,** measured in points from **baseline** to top. The depth of **descenders** is not included in this calculation. In some fonts the cap height is less than the **ascender** height of the lowercase letters, as in Bembo. The cap height should not be confused with the point size of a font, which includes the lowercase ascenders and descenders plus an additional space below. *See also* **Appendix 1.**

Capital figure. *See* **Lining figure.**

Capitals (caps). CAPITAL LETTERS, as distinct from **lowercase** or SMALL CAPITALS. *See also* **Small caps.**

Caps. Abbreviation for full-sized **capitals** or **uppercase.** *See also* **Appendix 8.**

Caps and lowercase (Clc, C/lc, C&lc). Instruction to set copy in a combination of capital and lowercase letters of the same type font, as in book and chapter titles (e.g., Five Ways to Boil an Egg); also called *upper- and lowercase* (Ulc, U/lc, U&lc). *See also* **Initial cap and lowercase; Appendix 8.**

Caps and small caps (Csc, C/sc, C&sc). Instruction to set copy in a combination of full and small capital letters of the same font; often used in primary subheadings (e.g., THE FIRST WAY TO BOIL AN EGG). *See also* **Small caps; Appendix 8.**

Caption. Originally used to mean specifically the title of an **illustration,** traditionally placed above it but now often placed below and followed by the **legend;** the two terms are often used interchangeably.

Carding. Also called *feathering;* the insertion of space between lines of type, in addition to the established **leading,** to fill out the depth of a page or column.

The word originated in the days of metal type, when strips of card or paper were used to insert the space. The practice was brought into disrepute by the desperate excesses of journal printers, who would use any device to accommodate late copy changes and still keep their pages square. Some photocomposition systems allow for the addition or subtraction of line spacing (known as plus or minus carding / carding out or carding in). The skilled typesetter has the capacity to manipulate the page depth with a subtlety that evades detection. Discretion is called for, however, and the practice remains controversial. RE

Caret (∧). Symbol used in manuscript or proof to mark the point where a correction or addition is to be inserted. *See also* **Appendix 8.**

Caron. *See* **Haček.**

Carry forward, carry over. (1) An instruction to move a text element down to the next line or over to the next page (*see also* **Appendix 8**). (2) In newspapers and magazines, a *carryover line* identifies the resumption of an article that has been broken over several pages.

Carry queries. An instruction to take unresolved **queries** forward to the next set of proofs.

Case. Originally the sectioned tray in which metal type was stored. Since capital letters were kept in the upper compartments and small letters in the lower, capital letters were called **uppercase** and small letters were called **lowercase.** During the nineteenth century in the United States the California case was introduced, dividing capitals and small letters right and left rather than top and bottom, but the established terminology survived this change.

Case fraction. A term (from the days of metal handsetting) for a **fraction** that is a single piece of type, housed in the same **case** as the other characters in the **font** rather than in a separate pi case. *See also* **Fraction; Pi character.**

Castoff, casting off. The calculation of the number of typeset pages a manuscript will make, based on a **character count.** *See also* **Copyfitting.**

Traditionally, the typesetter's use of the term assumes a very close calculation, accurate to within four to eight pages of the printed book length. This may take several days to prepare. Today many tend to use the term to describe a less precise calculation based on the average character count of a few manuscript pages. This is usually accurate to within 10 percent of the final page count and should be described as an estimate rather than a castoff. What is intended should be clearly understood, since a typesetter's castoff may cost several hundred dollars whereas an estimated page count will usually be prepared for little or no fee. CME/RE

Cataloging-in-Publication Data. *See* **CIP.**

Cedilla (˛). Diacritic used with consonants – with the letter c in French, Portuguese, and Catalan; with s and c in Turkish; and with s and t in Romanian. Not to be confused with the nasal hook or **ogonek,** which is used with vowels. *See also* **Appendix 7.**

Centered dot. A raised period centered on the **x-height.** Used in mathematical copy as a multiplication sign; also used as an ornament or for typographical emphasis. *See also* **Bullet.**

Centered typography. A style characteristic of traditional, symmetrical design, in which type elements are vertically aligned on a central axis, like the skeleton of a Dover sole.

Center on longest line. An instruction to move a **block** of type that has been set **flush** left, **ragged** right, so that the longest line is centered on the text measure.

Used most commonly in connection with verse and verse extracts to achieve a balanced layout, though the result is often the reverse. Alternatively, one may ask that an element be "visually centered"; that is, positioned so that it "looks right" – rather like hanging a painting with a warped frame, and equally frustrating. RE

Chapter head. Identification of a chapter that includes number, title, or both. Usually set off in larger or **display type**. *See also* **CT.**

Character. A single letter, numeral, punctuation mark, or any other element of a **font**. Also called **sort.**

Character count. (1) The total number of characters in a unit of measurement – a line, paragraph, or page. (2) The total number of characters in a manuscript. *See also* **Castoff, casting off; Copyfitting.**

In the days when manuscripts were typewritten, counting characters was easier because each character (or keystroke) occupied exactly the same amount of space. Manuscripts prepared using letters of proportional width make accurate character counts difficult. RH

Character set. (1) The characters in a given **font**. (2) The horizontal measurement of a letter, expressed in units of an **em**. (3) The characteristic width of a font. The character set of a particular type design may have several variants, that is, alphabets of the same height but of different widths; for example Helvetica, Helvetica Condensed, **Helvetica Expanded**. *See also* **Set, setting; Set width.**

Characters per pica (CPP). A measurement of the average number of typeset characters of any given font and point size that will fit in one pica.

C-head. The order of **subhead** third in importance, after **A-head** and **B-head**. *See also* **Appendix 4.**

Check repro. Instruction to the typesetter to check that an apparent scratch, blemish, or distortion on a proof does not appear on the **camera-ready copy.** Most such blemishes are introduced in the copying process, but this should never be assumed.

Cicero. A unit of type measurement in the **Didot point** system, which originated in France and is now employed throughout Europe except in the United Kingdom and Ireland. One cicero equals 12 Didot points, equivalent to 12.839 points in the Anglo-American **point** system. Ciceros are used in Europe to describe the width and depth of a block of text, serving much the same purpose as the Anglo-American **pica.**

CIP. Abbreviation for *cataloging-in-publication,* short for "Library of Congress Cataloging-in-Publication Data." This information is supposed to be reproduced exactly as provided by the Library of Congress; it usually appears on the **copyright page.** Sometimes it is redesigned or moved to the back of the book. *See also* **LC number; Appendix 2.**

Circumflex (^). *See* **Accent; Appendix 7.**

Clc, C/lc, C&lc. *See* **Caps and lowercase.**

Clean. (1) Used to describe a manuscript that is a sharp, clear copy with minimal editing. Since the condition of a manuscript is a factor in estimating typesetting costs, one that is heavily or carelessly corrected is called **penalty copy** and usually incurs an additional charge. Also used for proof or repro that requires no correction. (2) Instruction to touch out any blemish or **hickey.**

Clear. Instruction that specifies the width of the indention in **flush-and-hang** setting, used with particular reference to **endnotes,** when the note numbers are set to the left and the rest of the paragraph is indented. Thus, if the number of notes runs into the hundreds, the instruction might be, "Clear for 3 digits, period, and space. Align on period." For example:

I. Alexis Canote was later to establish his own atelier in a converted hunting lodge in the Bois.

I I. Lermontov was reported by his own second to have fired into the air, though his adversary denied this and even claimed to have been grazed on the cheek by the ball.

101. Madame Poincaré's self-medication turned all her fingernails black. It was for this reason that she painted them with red lacquer, and not through vanity as M. LeDuc would have us believe. Her "affected" laughter may be similarly explained, for mercury poisoning is known to cause a benign atrophy of the esophagus.

Clear text. Copy that replicates the style of the source document without "silent" editorial intervention.

Close spacing. An inexact instruction to use the minimum of word spacing consistent with legibility. The definition of "close" is subjective and can vary depending on the typeface. Helvetica, for example, may be more tightly word spaced than Gill Sans because the letters form tighter word patterns. **Letterspacing** may also be specified as close or tight, particularly in **display** setting. *See also* **Spacebands; Word space.**

Close up. An instruction in proofreading to delete space. *See also* **Appendix 8.**

Club line. A line of type that is considered aesthetically ill matched to its context; for example, a line that is shorter than a following **paragraph indent.** *See also* **Bad break; Orphan; Widow.**

CN. Abbreviation for *chapter number,* used in marking up manuscript. *See also* **Appendix 4.**

Colophon. From the Greek *kolophon,* meaning a finishing touch or flourish. Originally the printer's/publisher's title and device, placed on the last page of the book. This information is now included on the title page. In contemporary usage the colophon is a note about the production of the book, usually placed on the last recto or verso page but sometimes appearing on the copyright page. The colophon may list the typeface, paper stock, binding materials, number of copies (signed and unsigned), date and place of manufacture, identity of the typesetter, printer, and designer, and so on. The colophon is of particular interest to bibliophiles and is most commonly found in limited editions, but a number of scholarly and commercial presses regularly include a note on the typeface and other production information. *See also* **Imprint.**

Color. The visual **density** or texture of the text page. The color of a block of type is the result of many factors – primarily the typeface, but also type size, leading, margins, type measure, and word spacing.

A page is said to have a good color if it forms an even mass of gray. Squint at the page, and you will see this. A poorly designed typeface will look spotty, either because the individual letters are not of even weight or because their spatial relationship is erratic. If the word spacing is gappy, you will notice vertical streaks of white, known as rivers. A page that is inadequately leaded will appear dense and impenetrable. An overleaded page, or one with inadequate margins, will seem insipid and shapeless. RE

A recurring complaint is that photoset or digitized typefaces lack the color of the metal fonts from which they are derived. In letterpress printing the image spreads as the type is pressed into the paper, a circumstance that informed the design of the type. Many modern versions of classic book faces have not been adapted for lithographic printing, in which there is virtually no ink spread. RE

Column. A block of type.

Column head. A heading in a **table** that applies to only one column. *See also* **Decked head; Spanner head; Appendix 6.**

Column inch. A unit of measurement one column wide by one inch deep, used in ordering and pricing newspaper advertising space.

Column space. *See* **Ditch.**

Comp. *See* **Compositor; Comprehensive.**

Compose. To set type and make up **pages.**

Composition. The process of assembling characters, words, lines, and other blocks of type or pages for reproduction. This operation can be performed on a composing stick, on a **linecaster,** or with the aid of a computer program.

Compositor. Often abbreviated *comp.* A craftsman responsible for transforming manuscript into type, following a **house style** or a designer's or publisher's **layouts** and **specifications**. Compositors establish the parameters of the pages and their sequence and the relationship of all text elements, resolve any conflicts of style, and banish **widows, orphans, bad breaks,** and any other infelicities. *See also* **Keyboard; Typesetter; Typographer.**

Compound rule. *See* **Double rule.**

Comprehensive. Often abbreviated *comp.* A highly finished rendering of a **layout,** as close as possible to the printed piece, that the designer submits for approval before producing the **mechanical.** Used most commonly in publishing with reference to **dust jacket** or cover designs. Sometimes called a *finished rough* or *tight comp.*

Computer generated. Term used most commonly for elements created or modified by the computer because they are not included among the resident **fonts.** For example, **small caps** may be simulated by the computer if **true-cut** small caps are not available. *See also* **Fake small caps; Machine italic; Small caps.**

Condensed type. A narrow **typeface,** usually part of a type **family** that also includes a version of regular width. Condensed (and expanded) faces may be either "cut" (original, designed fonts) or **computer generated.**

Consecutive hyphens. End-of-line hyphens in consecutive lines of type; also called *ladder.* In the United States it is generally considered unacceptable to have more than three hyphens in a row in **justified** composition, or more than two in **ragged composition.** The typesetter is expected to rebreak the lines as necessary to remain within this limit, and sometimes loose word spacing results. *See also* **Break block; Hyphen block.**

This horror of hyphens is a relatively modern phobia. As recently as the nineteenth century, the priorities were reversed: clusters of five or more hyphens were common, and the word spacing was uniformly tight. RE

Contents. A listing of elements of the work, part of the **front matter.** *See also* **Appendix 2.**

Continuous-tone art. Art with gradations of gray to be reproduced as a **halftone.** Usually shown in typeset pages as **FPO.** *See also* **Line art.**

Copy. (1) **Manuscript** that is to be typeset. (2) Type or other **artwork** that is ready to be photographed by the platemaker, as in **camera-ready copy.**

Copy block. A sequence of lines of type treated as a single element in design or **page makeup.**

Copyediting. Marking a manuscript for correct usage – including spelling, grammar, syntax, and punctuation – and for consistency and conformity to the publisher's **house style,** as opposed to editing for content. The copyeditor also marks the manuscript to identify such elements as extract and subheadings, notes any queries that require the author's attention, and compiles a **style sheet** as a guide to the typesetter. Also called *manuscript editing. See also* **Appendix 4.**

Many copyeditors, especially at university presses, also do a considerable amount of substantive editing, making suggestions for cutting, expanding, rewriting, or reorganizing as needed. N B P

Copyfitting. (1) Design and choice of type. (2) Editing to make copy fit into a prescribed space. Often used as a synonym for **castoff.**

Copy preparation. (1) Checking and correcting a **manuscript** before typesetting to ensure that the fewest possible errors occur in the first **proof; copyediting.** (2) Marking **camera-ready copy,** such as photographs, with instructions to the printer regarding the percentage of enlargement or reduction, **cropping,** positioning, screen gauge, and so on.

Copyright page. The page that carries the copyright information and the **International Standard Book Number,** usually the **verso** of the **title page.** May also include **CIP,** place of manufacture, number of the edition and impression, **acknowledgments** of subventions, the **colophon,** and the author's **dedication.** *See also* **Appendix 2.**

Corner marks. (1) Marks inserted by the compositor in page proof to define the **type page,** often in the form of two fine horizontal rules indicating the baseline and cap height of the topmost element on a type page, or in the form of crosshairs indicating the topmost element on the type page. Also called *stripping guides.* These marks are required on any page where the standard **margins** are not obviously defined by the text block itself, as is often the case with **front matter,** chapter openings, and pages of illustrations. The printer uses them as a guide when assembling the pages before platemaking and deletes them once they have served this purpose. Not to be confused with trim marks, which indicate the final size of the page. (2) Four L-shaped marks that indicate the corners of the area of a **halftone** or other graphic element to be added to the page later. *See also* **Register marks.**

Corrections. (1) Changes made in the manuscript by the author or editor. (2) Changes made in typeset proof (*see* **Alteration**). (3) Typeset proof, incorporating any corrections made in the previous proof. These may be supplied as completely rerun galleys or pages or as patch or **strip-in corrections,** blocks of a few lines to be pasted over the altered lines in the previous proof. In either case, care should be used to ensure that the **density** of the correction matches that of the earlier setting.

Corrigendum (plural *corrigenda*). An error, especially a printer's error, discovered too late to be corrected within a document and included in a list printed in a book; sometimes used interchangeably with **erratum.** *See also* **Addendum.**

Counter. A partially or fully enclosed white space that forms part of a letter, such as the doughnut hole of the letter O. *See also* **Appendix 1.**

Coupon box. A box shape defined by a **coupon rule,** often used in newspapers and magazines to indicate the portion of an advertisement that the reader is to cut out and return to the advertiser. Sometimes a scissors icon or other typographic device is included to make the intention plain.

Coupon rule. A rule made up of spaced dashes. Since the weight and length of dashes and the amount of white space are variable, it is best to provide the typesetter with **specifications** or a sample.

CPP. *See* Characters per pica.

CRC. *See* Camera-ready copy.

Crop. To omit part of an image when converting **artwork** to its final size. The area to be cut away, or *cropped,* may be indicated by *crop marks* (usually vertical and horizontal lines in the margins of the **original**), by a rectangle on a tracing paper overlay showing the portion to be used, or on the back of the image.

Crop marks. *See* Crop.

Crossbar. A horizontal stroke that crosses to, over, or through the **stem** of a letter; may be used as a diacritic, as in capital edh or Hausa Đ. Also called *hairline bar* in some faces. *See also* **Ligature; Appendix 1; Appendix 7.**

Crosshairs. *See* **Corner marks; Register marks.**

Crosshead. *See* **Spanner head.**

Cross rule. *See* **Spanner rule.**

Csc, C/sc, C&sc. *See* **Caps and small caps.**

CT. Abbreviation used in marking manuscript and proof to identify a heading as a chapter title. If other elements in the text, such as the headings for the contents page and acknowledgments, are to be set in the same typographic style as the chapter titles, they should be marked in the same way. *See also* **Appendix 4; Appendix 8.**

Ctr. Abbreviation used in marking manuscript or proof to indicate that an element should be centered with reference to the text width or another specific element on the page. *See also* **Appendix 4; Appendix 8.**

Curly brace. *See* **Brace.**

Curly brackets. *See* **Brackets.**

Cursive. Descriptive of a **typeface** that employs the flowing strokes of handwriting but in which letters are separate. Though **italic** shares this characteristic, it is not categorized as cursive because of its specific role as a complement to **roman.** *See also* **Script.**

Cut. A specific version of a **typeface,** derived from the cutting of punches for the production of metal type. In metal type, the cut of the face varies from one size to another. A separate set of punches must be prepared for each, and subtle variations of stroke weight, **counter** size, and **x-height** may be introduced to maintain legibility and consistency of **color**. In most photosetting systems this enhancement is lacking because all type sizes are photographically derived from a single master, usually the 8-point or 12-point **font**. *Cut* is also used to distinguish between versions of the same typeface available from different manufacturers. For example, there are many cuts of Garamond available, varying in their faithfulness to the original.

Cut and paste. *See* **Pasteup.**

Cut-in head. (1) A heading set into an indented area of the text. (2) A subhead in a table that begins in the **stub** but projects into other columns.

Cut-in initial. *See* **Drop cap.**

Cutting in. *See* **Mortise** (2).

Cut to center. Instruction in marking manuscript or proof to indicate that a text element should be moved to center on the width of the text **measure** when pages are made up; used most commonly for poetry **extracts,** where the relation of the verse lines to the text above and below depends on where page breaks occur. *See also* **Appendix 4; Appendix 8.**

Cut type. *See* **Broken type.**

Cyanotype. *See* **Blueline, blue.**

Cyrillic. Designating the Old Slavic alphabet, derived from Greek **uncials** and attributed to Saint Cyril; used in modified form to write Russian, Bulgarian, and some other languages of the former USSR.

DA. Designer's alteration. *See also* **Alteration.**

Dagger (†). Typographic symbol used to indicate an unnumbered footnote. *See also* **Reference mark.**

Dash. A short typographic rule used for various purposes. Common lengths are: en (–), em (—), 2-em (——), and 3-em (———). Of these the most frequently used in the United States is the **em dash**, to indicate a break in a sentence, an interpolation, etc. British practice, also favored by some U.S. typographers, is to use a word-spaced **en dash** for these purposes. A possible compromise between these two preferences is a new character, a ¾-em dash centered in an em space. The 3-em dash is used in bibliographies to stand for the name of the author or authors cited immediately above. Some consider this unsightly and prefer to repeat the names. The **hyphen** is usually not considered a dash and frequently has a different weight (thickness). *See also* **Em dash; En dash.**

Dashed rule. *See* **Coupon rule.**

Dead copy, dead matter. Manuscript or proof that has been superseded by corrected typeset proofs and thus is at least two stages back in the production cycle. Often used interchangeably with **foul copy** and **foul proof,** although strictly speaking the latter are not "dead," since they may be referred to again, usually to distinguish between **AAs** and **PEs.**

Decked head. A heading with **rules** above and below; especially a **subhead** within a **table** body, with rules above and below that either run the full table measure or span the columns to which the subhead applies. *See also* **Appendix 6.**

Decorative initial. A large initial character, embellished with scrollwork or other ornamentation and often shaped to a rectangle, the better to align with the following text. Decorative initials, derived from early Christian manuscripts, have not generally been available in photocomposition, but this is changing. *See also* **Drop cap; Raised initial.**

Decorative type. *See* **Appendix 3.**

Dedication. Author's inscription, formerly often thanking a patron for favors received or anticipated, now customarily expressing esteem or affection for the person or persons to whom it is addressed. If the author wishes to include a dedication, it is usually placed on a recto page following the copyright information, but if space is limited it may appear on the copyright page. *See also* **Appendix 2.**

Delete. Instruction on manuscript or proof to omit material marked. *See also* **Appendix 8.**

Density. The blackness and especially the "thickness" or "weight" of set type. A compositor should be able to provide type of uniform density throughout a book (appropriate to the particular typefaces employed), including all corrections and regardless of whether the corrections are set as complete pages or as **strip-in corrections.** *See also* **Color; Output resolution; Resolution.**

Depth of page. The measured vertical dimension of a page, calculated by different typesetters in different ways. *See also* **Text page; Trim; Type page.**

One of the most common complaints of compositors and layout artists is that apparently precise page specifications, such as "44 picas 3 points," do not agree with the material to be set on a page. Another complaint is that the depth-of-page specification may be unclear about which elements are included: does "44 picas 3 points" refer to the base-line of the last text line or to the descenders of the last text line? This matter is of particular concern when pagination is done on the computer rather than the light table, especially when the design allows vertical spacing to vary in order to align spreads. Under these conditions, the compositor must program the information about vertical space and its tolerances just as precisely as the information about horizontal space (measure, indents, etc.) and its tolerances (word-space parameters). The wise designer is either very careful to be precise or avoids the points-picas specification by giving page depth in some other way (e.g., the head margin, the space from running head to first text line, and the number of text lines). RA

Descender. The portion of a **lowercase character** that falls below the **baseline** or **x-height,** as in g and p. Descenders also occur in **old-style figures** 3, 4, 5, 7, 9. *See also* **Long descender; Short descender; Appendix 1.**

Designer's alteration (DA). *See* **Alteration.**

Desktop composition. A popular generic term describing a range of microcomputer-based typesetting systems that output **camera-ready** pages on a wide range of **imagesetters**. Desktop composition is commonly, but erroneously, called *desktop publishing.*

D-head. An order of **subhead** subordinate to A-, B-, and C-level sub-heads.

Diacritic, diacritical mark. Mark placed over, under, or through a letter to distinguish it from one of similar form or to indicate pronunciation or stress; often used interchangeably with **accent.** *See also* **Appendix 7.**

Diaeresis, dieresis (¨). Diacritic positioned above the second of two successive vowels to indicate its pronunciation as a separate syllable, as in *naïve. See also* **Umlaut; Appendix 7.**

Diagonal. *See* **Slash.**

Diamond. Several different character shapes with four sides, either solid or outlined, and including a turned square; may be used as a diamond **ornament.** When a particular shape is wanted, an example should be included. *See also* **Lozenge.**

Didot point (d). A typographic measurement introduced in 1775 by the French printer François-Ambroise Didot and used in continental Europe. The Didot point equals 0.375 mm or 0.015 inch, larger than the Anglo-American **point;** for example, 12 d is approximately the same as 14 points in the Anglo-American point system. *See also* **Cicero.**

Digit. (1) Any individual **arabic numeral,** from 0 through 9. (2) A typographic symbol (☞) used to "point out" a specific element; also called a *printer's fist* or *index,* the digit was once included in the **case** with other standard **reference marks.**

Digitized characters. Type character information stored in digital (electronic) form. Earlier methods of type composition used characters made of metal or character patterns on photographic film negatives.

Dingbat. *See* **Ornament.**

Diphthong. In typography, the combination of two vowels into a single character and a single sound, Æ, æ, Œ, œ. *See also* **Ligature.**

Dirt. A general term for spots, scratches, blemishes, or stains on proofs or printed material that need to be brought to the typesetter's attention.

Discretionary hyphen. *See* **Soft hyphen.**

Disk-to-film. A process by which electronic files, containing all necessary information for imaging, are taken directly from their storage disks and RIP'd (raster image processed) onto film without any intervening **repro** stage or camera work. The resulting film may either be generated as loose single pages or preimposed into correct position. *See also* **Imagesetter.**

Because there is no repro stage, it is extremely important that all corrections be made in last page proof. With this process it is also imperative that the designer, from the very beginning of the job, makes sure to keep in close communication with typesetter and printer to ensure that all systems and requirements are possible and compatible. AWS

Disk-to-film is a current buzzword; electronic prepress probably better describes the technique. CME

Display. (1) To treat certain elements of a manuscript, such as the title page, chapter titles, and subheads, differently from the body text. (2) **Display type.** *See also* **Display face; Titling font.**

Mathematical expressions or a single line of verse may also need to be displayed. For example, an author may have been able to run an equation into the text line in a double-spaced typescript, but because of its size or structure an editor may mark it "display" for typesetting. CME

Display face. A typeface designed for use in large sizes rather than for body text. A type **family** may include a display font as well as a text font; the boldness and stroke weight of the display font will differ from those of the text font. In computer composition, a font should be called a "display font" only if it is set from a master (usually an 18-point master) different from the one used for the text (usually a 12-point master). Some typefaces designed exclusively for display setting are limited to **capitals** and **lining figures** (also called **titling fonts**); for example, Michelangelo.

Display type. (1) **Display face.** (2) The fonts used for the title page, chapter openings, subheads, and other elements different from the body text. *See also* **Appendix 3.**

Ditch. The space between two columns of text, lists, etc. The term is not generally used when specifying space in indexes or tables. *See also* **Gutter** (2).

Dot leader. *See* **Leader.**

Dotless i (ı). A lowercase i without the dot, used in Turkish; also used when an accent is to be "floated" over the i, as in the Irish **faíthe,** and occasionally with f **kerned** to simulate an fı **ligature** if the true sort is not available. Also called *undotted i.*

Dotless j (ȷ). A lowercase j without the dot, used when an accent is to be "floated" over the j, and occasionally with f **kerned** to simulate an fȷ **ligature.** Encountered most often in northern European languages. Also called *undotted j.*

Dots per inch (dpi). (1) The frequency of dots in rendering **digitized** images or type characters with a laser **imagesetter;** 300–600 dpi imagesetters are usually referred to as low **resolution,** 800–1,000 as medium resolution, and 1,100 and above as high resolution. (2) The frequency of **halftone** screen dots within a linear inch. Halftone dots are created using ruled screens, which are referred to by the number of lines per inch (lpi) in the screen. *See also* **Output resolution.**

Double column. The arrangement of text elements into two narrow type blocks of specified width placed side by side on a page. The instruction "short notes are to be double-columned" means that a succession of a certain number of footnotes or endnotes, each less than half a page wide, should be set in two columns on the horizontal measure even though the general style is for notes to be set in a single column.

Double dagger (‡). Typographic symbol used to indicate an unnumbered footnote. *See also* **Reference mark.**

Double keyboarding. In composition, a method of ensuring accuracy by keying in a manuscript twice, comparing the two versions by computer, and resolving any differences. The advantage for the compositor is that double-keyed **proof** is usually cleaner than proof that is keyed in and proofread. The primary disadvantage is that double keyboarding is slower than single **keyboarding** and **proofreading**.

Double-page spread. *See* **Spread.**

Double quotes. *See* **Quotation marks.**

Double rule. Two **rules** running parallel to one another. Since there is no standard double rule, the weight, length, and spacing must be specified. Also called *compound rule. See also* **French rule; Scotch rule.**

dpi. *See* **Dots per inch.**

Drop. (1) To measure vertically down the page, as in "drop 16 points to base" (*see also* **Sink**). (2) To omit or delete an element.

Drop cap. An **uppercase** character set in a type size larger than the text and "dropped" or extended into lines of text, usually at a chapter opening and usually **base aligned** with the second or third text line. The specification for a drop cap should include the way the adjacent text lines should be set. The three most common styles are: all text lines **block** or **box indented**; first text line set close, subsequent lines block indented; and all lines shaped or run around the cap. Also called *cut-in initial, drop initial, sink* or *sunken initial,* or *initial letter* and contrasted with **raised initial.** *See also* **Decorative initial; Runaround.**

Drop folio. A page number positioned below the normal **text page** on chapter openings or other pages when all other folios throughout the book are at the top or some position other than the bottom. Some consider a drop folio to be within the **type page;** properly speaking, it should be considered outside the type page. Not to be confused with **foot folio.** *See also* **Bounce; bouncing folio.**

Drop initial. *See* **Drop cap.**

Dropped folio. *See* **Drop folio.**

Duck feet. *See* **Guillemets.**

Dummy. Manual **pasteup** of **galley proofs** of a book, or a computer-generated version, with all the visual elements – **text** and art-

work – in place. The dummy is usually made by the designer or under the designer's supervision and used as a guide for preparing pages or **camera-ready mechanicals.** The **layout** may be somewhat rough, but the dummy must be accurate enough so that, when pages are made or **repro** is pasted up, column breaks, page breaks, and artwork will fall as shown on the dummy.

Dust jacket. Printed paper covering wrapped around the binding of a casebound book. Originally intended to protect the book, the dust jacket is now used principally as a marketing tool.

There are opposed schools of thought about the relation of jacket and text design. Scholarly publishers have tended to favor a definite correspondence between the two, whereas trade houses stress the jacket's independence as a miniature billboard. This difference of approach is partly a reflection of the various markets that scholarly and trade houses serve. It has to be admitted, however, that university press jackets often appear constrained and tentative. Two of the greatest modern authorities on book design, Jan Tschichold and Stanley Morison, saw the functions of jacket and book as quite separate. Indeed, Tschichold has said that the first thing to do when one takes a book home is rip off the jacket to display the binding. RE

Dutch Old Style. Name given to seventeenth- and eighteenth-century type designs used by Dutch printers. These were based on earlier models of Claude Garamond's punches collected by Christophe Plantin of Antwerp. The Dutch designs later influenced English type designs, one obvious example being Caslon. Examples of Dutch Old Style that survive (in some form) include Erhardt and Janson. *See also* **Appendix 3** (old face).

E ·

EA. Editor's alteration. *See also* **Alteration.**

Ear. A short, rounded **terminal,** as on the lowercase roman g of most book types. *See also* **Appendix 1.**

Edh, eth (Ð, ð). Character used in **Old English** and Icelandic to represent the voiced th.

Edited MS. A manuscript that has been fully **copyedited** and is ready for production (design and composition).

If, to save time, an unedited manuscript is passed to production (as inevitably happens now and then), subsequent editorial changes may lead to considerable delay and frustration. RE

A further problem with manuscripts that are designed before copyediting is completed is that some things may be marked by the editor in a manner that conflicts with design specifications – such as when both acronyms and abbreviations are marked by the editor to be set in full caps (or small caps) when the designer has specified small caps for acronyms and full caps for abbreviations (or has chosen a typeface without small caps). These confusions are passed on to the compositor, who frequently has to pass them back to the publisher in first proof. When this happens, second proof typically takes longer than normal, and alteration costs are inevitably higher. RA, RE, CME, RH, NBP, AWS

Editing. *See* **Copyediting.**

Editorially correct disk. Computer file into which all editorial changes to the author's manuscript have been made. Sometimes called *updated disk.*

It is not always easy to specify exactly what is meant by an editorially correct disk; for example, if the A-heads are specified by the designer to be set upper- and lowercase but are full capitals in the files, are the files "editorially correct"? CME

Editor's alteration (EA). *See* **Alteration.**

Egyptian. Family of **typefaces** of early nineteenth-century origin in which strokes and **serifs** are of equal or similar weight. Also called *slab serif. See also* **Appendix 3.**

Electronic manuscript. Manuscript stored on magnetic media, such as tape or floppy disk, as distinguished from **hard copy;** may include **generic codes** as well as text. *See also* **Editorially correct disk.**

Elite type. The smaller of the two common U.S. monospace typewriter faces, which gives 12 **characters** to the inch. Though less common, machines of 11, 13, and 15 characters per inch have also been manufactured. *See also* **Pica type.**

Ellipsis points. Three evenly spaced periods (. . .) used to indicate the omission of words within a quoted text or a deliberately incomplete sentence. *The Chicago Manual of Style* specifies a 3-to-em space, and some typographers feel the space should be equal to the **word space,** varying like the other word spaces in a **justified** line. PostScript fonts usually come with an ellipsis that has a fixed space between the periods – one considerably smaller than the 3-to-em space. *See also* **Poetry ellipsis.**

In general, it is preferable not to have ellipsis points begin a new line, but if avoiding this results in excessive word spacing or awkward hyphenation, the typesetter may choose to break a line before the ellipsis. Except in very rare cases, it is not permissible to have ellipsis points stand alone as the last line of a paragraph. CME

Em. In current usage, a unit of measurement that equals the **point** size of the type. Thus the value of an em space when setting 8-point type remains 8 points, regardless of the **leading.** As a linear measurement the **em** (or fractions thereof) is used to specify indention, spaces within the line, and the width of elements such as **dashes.** The em is the basis for word spacing **justified** type and for the **unit** system. An em space is marked on the manuscript by a square box (□) or a square box with the letter M inside. Multiple ems may be marked with a box containing the number of ems, as ②, ③, etc. *See also* **Word space.**

Em dash. Typographic rule usually measuring the width of an **em.** Usually 2- and 3-em dashes are set by keying two or three em dashes in a row. Some fonts, however, have an em dash considerably shorter than an em space, with the **set width** of the character remaining an em. *See also* **Dash.**

Em quad. A unit of measurement the square of the type size. In metal setting, a blank of this size, used as spacing material. So called because the capital M was usually cast on a square **body.** *See also* **Em; Quad.**

Em space. *See* **Em.**

En. A linear typographic measurement half the width of an **em.** The en space is marked on the manuscript by a half-filled or divided box (◪ or ◩). *See also* **Em; Nut.**

En dash. Typographic rule measuring (more or less) the width of an en space. In many typefaces where the figures occupy more than half an em space, the en dash has the same width as the figures – that is, slightly more than an en space. In the United States the en dash is used with numbers to mean "through," for example, pages 12–24; a hyphen is used where "through" is not implied, as in telephone numbers or **International Standard Book Numbers.** The en dash is also used in compound adjectives where one element is two words or a hyphenated word. In the United Kingdom, a word-spaced en dash is used in place of the **em dash,** a practice favored by some U.S. typographers who believe it is visually less disrupting in a line of type. *See also* **Dash.**

End matter. *See* **Back matter.**

Endnote. Numbered note to the text placed at the end of the chapter it refers to or at the end of the book, rather than at the foot of the page where the text reference occurs. *See also* **Footnote.**

Eng (ŋ, ŋ). Character used to represent the sound *ng* as in *thing.*

En space. *See* **En.**

Epigraph. Quotation of prose or verse set on a page in the **front matter** to introduce a book, following the chapter title to introduce a chapter or section, or within the chapter following a subhead or subsection. *See also* **Appendix 2.**

Equalize space. Instruction marked on proof to make two or more (visually) unequal spaces visually equal. *See also* **Appendix 8.**

Erratum (plural *errata*). A correction to material within a document included in a list that is printed separately and either tipped in or inserted loose inside the front cover of the book; sometimes used interchangeably with **corrigendum.** *See also* **Addendum.**

esc. *See* **Even small caps.**

Escapement. On a typewriter, the movement of the carriage after a **character** or spacing key has been struck. On monospacing typewriters the escapement is always the same. On proportional-spacing typewriters the escapement varies depending on the key (character) struck. The term has been carried over to describe the behavior of machines that do not have a carriage, such as many typesetting machines. It is usually used to describe a machine function rather than the properties of a character, although conceptually the escapement for a typesettable character would be the space equal to the character plus its left and right **sidebearings.**

Eszett (ß). Character sometimes used in German for ss.

Eth. *See* **Edh.**

Ethel (Œ, œ). **Ligature** formerly used in English and still used in French.

Even small caps (esc). All **small caps,** sometimes indicated by just *sc.* Some designers use this term to indicate that small capitals should be "gracefully" letterspaced, with the amount of **letterspacing** either indicated in **units** or **points** or left up to the compositor. *See also* **Appendix 8.**

Exception dictionary. Some typesetting systems (programs) hyphenate words according to an algorithm rather than a stored dictionary. Since no algorithm has been devised that can adequately hyphenate English, many algorithmic **hyphenation** programs also provide for a small "exception dictionary" into which the user can enter words and their proper hyphenation points. The program's word-division routine will be set up so that the exception dictionary is consulted first, and only when a word is not found there will the algorithm be applied.

Ex-height. *See* **x-height.**

Expanded type. A wide **typeface,** usually part of a type **family** that includes a version of regular width; sometimes called *extended type.* As with **condensed type,** the **font** may be either designed in this version or **computer generated.**

Expert font. A supplement font to a PostScript **typeface family** that contains, among other things, **small caps, old-style figures,** and true-cut **superscript** and **subscript** figures. Generally, small capitals are offered only in the **roman** fonts and in a few **bold** fonts.

The Linotype supplemental fonts have small capitals and old-style figures in the roman font, and old-style figures in the other fonts, but do not have true-cut superior or inferior figures, double-f ligatures, etc. CME

Extended type. *See* **Expanded type.**

Extra bold. A **typeface,** usually part of a type **family,** that is thicker or heavier in weight than the **bold** design.

Extract. Quoted matter, prose or poetry, set off from the text by **displaying** it rather than placing it in quotation marks. Extracts are differentiated from the author's text by various typographic techniques, such as a change in type size, a change of font, indention, extra space above and below, or a combination of these. Also called *block quotation.*

Extra lead. In metal composition, the addition of extra vertical space between typographic elements, usually specified in **points** in addition to the **leading** in effect, for example, "plus 6 points." The term may still be used for photocomposition, but to avoid ambiguity it is preferable to specify the extra space in a **base-to-base** measurement; for example, if 12-point leading is in effect, specify "18 pts b/b" between two elements rather than "plus 6 pts."

Eye. The inner, fully enclosed (**counter**) portion of the lowercase e. *See also* **Appendix 1.**

Face. Short form for **typeface.**

Face margin. *See* **Outside margin.**

Facing pages. Left (**verso**) and right (**recto**) **pages** opposite one another when a book is lying open. Also called **spread;** contrasted with **leaf.**

Faíthe. Accented i in Irish Gaelic. *See also* **Appendix 7.**

Fake small caps. Small caps created by electronically reducing full-sized capital letters in a font to approximately the **x-height.** To be most effective, capitals are not only set to a smaller size but also slightly **expanded.** However small caps are faked, they tend to lack the weight of **true-cut** small caps and can look thin in a line of text. Since most fonts do not have **italic** small caps, a passable version can sometimes be generated by slanting true-cut small caps. A few typefaces do have **bold** small capitals, but most bold small caps also have to be faked. *See also* **Machine italic.**

With PostScript fonts, it is possible to create some reasonably good fake small caps using programs such as Fontographer. Essentially the technique involves scaling the full caps nonproportionally, then increasing the stroke weight. For example, using Fontographer terminology, the caps may be scaled 66 percent vertically and 70 percent horizontally, and the stroke weight increased by 8 em units. Experimentation is needed with each font to determine the best values. This is not a perfect solution; for example, the proportioning of the serifs will always be a little off, the crossbar on the A will not be quite right, etc., but on balance the technique is useful. Further touching up of such fake small caps is possible, depending on the skill and patience of the person using the program. The final characters can then be written off as a Post-Script file for general use in composition. CME

False title. *See* **Bastard title.**

Family. The most comprehensive grouping of **typefaces** evolved from a common design, including all its variations of weight, width, size, and italics. *See also* **Appendix 3.**

Fat face. A typeface derived from modern letterforms and characterized by an extreme contrast between the thick and thin strokes.

Feathering. *See* **Carding.**

Fiducial box. A **box** having round corners.

Figure. (1) **Numeral.** (2) An **illustration** printed along with the text; also called *text figure;* contrast **plate.**

Figure space. A space equal to the space taken up by a **numeral** in a

font. In most **text fonts** all the numbers have the same **set width,** frequently, but not always, an **en** space; **display faces,** on the other hand, commonly have figures set on differing widths. The term is useful only when all figures have the same width.

File conversion. Converting a computer file created by specific software into a file readable by some other software in such a way that most or all of the **formatting** remains intact. *See also* **Media conversion.**

Fine rule. *See* **Hairline rule.**

Finial. *See* **Terminal** (2).

Finished rough. *See* **Comprehensive.**

First half title. *See* **Bastard title.**

First-order character. Full-sized **character.**

First proof. Traditionally, **galley proof** was the first proof returned by the compositor; more recently the first proof is often a rough **page proof,** and it is therefore clearer to refer to a first proof rather than a galley proof. *See also* **Rough pages.**

First revises. The second proof returned by the typesetter, including the corrections marked on the **first proof.** First revises may be pages if the first proof was **galley proof** or revised pages if the first proof was **page proof.** Infrequently, first revises may be revised galley proof if there were a large number of alterations to the first set of galley proof or if the galley proof will be used to create a paging **dummy.** *See also* **Revised proof.**

Fit. The spatial relationship of consecutive **characters,** as in the comment "the letter fit of ITC Garamond is too tight."

Fitting. *See* **Copyfitting.**

Fixed folio. A page number that never varies in position. *See also* **Bouncing folio.**

Fixed space. A space that does not vary in relation to its point size, unlike a **word space,** which may vary in **justified** copy. Fixed space is sometimes taken to have a specific value, such as ⅓ of an **em,** but this is a dangerous assumption. A designer who wants a specific horizontal space after a particular element – for example, following a **footnote call** – should specify the value in either **units** or **points.**

FL. Abbreviation for *flush left. See also* **Flush.**

Flap copy. *See* Jacket flap copy.

Fleuron. *See* **Flower.**

Float. (1) Instruction to center an element vertically and horizontally within a defined area on the page. (2) Instruction to the typesetter to send **repro** proof as it has been generated without placing it in position, as in "float the title page repro."

Flop. Instruction to turn an **illustration** or graphic image over so it faces the opposite way, either because it has inadvertently been reversed or because the designer prefers it.

Flourish. An embellishment or **ornament,** usually calligraphic or composed of graceful, curved lines.

Flower. A typographic **ornament** of floral design, such as a Granjon flower; also called *fleuron* or *printer's flower* (one example, ✿).

Flush. Aligned to a common edge, as in *flush left, flush right,* or *flush to gutter margin. See also* **Ragged composition.**

Flush and hang. A specification for setting text such as a bibliography, list, or dramatic extract with the first line flush left and all the following lines indented the same amount, as in this glossary. Also called *biblio style.*

Flush left (FL). *See* **Flush.**

Flush right (FR). *See* **Flush.**

FM. *See* **Front matter.**

FN. *See* **Footnote.**

Folio. Number indicating page sequence within a document, often used interchangeably with *page number.* Also used to refer to the page number when specifying a table of contents, as in "folios follow chapter titles after an em space." *See also* **Bouncing folio; Drop folio; Fixed folio; Foot folio.**

Follow copy. Instruction to preserve the spelling, punctuation, or any other element in the manuscript, even where this element is different or eccentric according to modern usage or the publisher's specified **house style.**

Follow style. Instruction to the typesetter to ignore what may seem a variation in a manuscript (such as a numbered list not typed **flush and hang**) and treat it according to the previously established style for that manuscript.

Font. A particular cut of a **typeface**, traditionally associated with one particular size, for example, 10-point Futura or 24-point Baskerville. Modern usage tends to interchange the definitions of *font* and *face*, because in photocomposition many sizes are photographic enlargements or reductions from a single master (**cut**). In metal type, each size of type is separately designed and cut. If more than one master is available for a particular face, each is considered a separate font.

Font metrics. Properties of **photocomposition** fonts, such as widths, heights, depths, and **kerning** pairs, usually stored as a separate file – the AFM (Adobe Font Metrics) file in PostScript fonts – that may be used by the composition program. The term *font* is used here without respect to final setting size, and the values of the program are in relative measurements, so that they apply to all sizes of type set from the master (the **font**). How well preset values work in setting type depends on a host of factors, such as the size of the type being set and whether it is computer modified (e.g., condensed or expanded).

Foot folio. Page number positioned at the bottom of the **type page** throughout the book. A folio placed at the bottom of a chapter opening page when all other folios throughout the book are at the head (top) of the page is called a **drop folio**. *See also* **Bouncing folio.**

Foot margin. *See* Bottom margin; Margin.

Footnote (FN). Note to the text placed at the bottom (or foot) of a page and referenced to the text by a number, letter, **asterisk, dagger,** or other device. *See also* **Endnote; Footnote call; Reference mark.**

Footnote call. The number or symbol used in a text, table, or illustration to indicate that there is a note to what is being discussed and to identify which note. There are certain general practices: (1) When numbers are used, they should be **superscript lining figures** even when an instruction says "use **old-style figures** throughout." (2) If superior figures are used for the note number (as well as the call), the size should be the same unless there are specific instructions to the contrary or unless there is more than a 2-point difference between text type and note type. The same rule applies when symbols such as the **asterisk** and the **dagger** are used for notes. (3) All note calls in a sequence are the same style of type. Table notes constitute a different sequence and are usually indicated by raised roman lowercase letters, so as not to be confused with either exponents or variables (*see also* **Appendix 6**). Should a designer or editor wish to vary from these practices, preference should be stated in the **specifications**. *See also* **Reference mark.**

Foot trim. *See* Trim, trim size (2).

Fore-edge margin. *See* Margin; Outside margin.

Fore-edge trim. *See* Trim, trim size (2).

Foreword. An introductory statement about a book usually written by someone other than the author and included in the **front matter.**

> *Extraordinary vigilance is required to prevent this from appearing as "forward."* NBP

Format, formatting. (1) The dimensions, layouts, and type specifications for a design. (2) Instructions (such as type size, measure, leading, and indentions), stored in a computer program, which can be called out either by a code embedded in the file or by a few keystrokes; similar to "styles" or "style sheets" used in many word-processing programs. (3) The size of a book page. Common formats, determined by standard paper and machine sizes, are $5\frac{1}{2} \times 8$, 6×9, 7×10, and $8\frac{1}{2} \times 11$ inches. (4) A standard generic design, such as is used for journals and some book series.

Foul copy, foul proof. **Manuscript** or **proof** that has been superseded or revised and is thus obsolete, though it may occasionally be referred to, usually to distinguish between author's **alterations** (AAs) and **printer's errors** (PEs). Often used interchangeably with **dead copy.**

Foundry type. Reusable metal type, purchased from the foundry as individual type **characters** rather than as molds. This type has usually been cast with a harder alloy than that used for **Monotype** or **Linotype** hot-metal type machines.

FPO. Abbreviation of *for position only.* A copy of a graphic element, also called a *position print,* used to indicate exact placement. It should be clearly labeled FPO with red or black ink to indicate that it is not copy for reproduction. FPO prints should be placed in as early a proof stage as is practical, so that the author, editor, and designer can check that the right image is in the right place and that size, orientation, and cropping are correct. *See also* **Blueline, blue.**

FR. Abbreviation for *flush right. See also* **Flush.**

Fraction. Specification of fractions is one of those matters complicated by changing technology; the old descriptions used with metal type seem somewhat useless today. There are three basic styles of setting fractions, and though various names are used to refer to the same style, we prefer the following terminology: (1) *online fraction* (1/2), in which the numerator and denominator are full-sized figures separated by a **slash** and both numbers rest on the **baseline;** (2) *piece fraction* (½), in which the numerator is a superior figure, the denominator is the same size but resting on the baseline, and the numbers are separated by a slash or *fraction bar;* (3) *case fraction* ($\frac{1}{2}$), which can either be taken from a generic casefraction font or created by using superior figures from the same font as the rest of the text, but with the numbers separated by a horizontal line instead of a diagonal.

Fraction bar. *See* **Slash.**

Fraktur. A form of **black-letter** design formerly used in German manuscripts and printing. *See also* **Appendix 3.**

French quotes. *See* **Guillemets.**

French rule. A double typographic **rule** divided by **diamond ornaments.** *See also* **Double rule.**

French spacing. Extra space added after the period at the end of a sentence. Common in typewritten manuscripts but considered redundant in typesetting.

In photocomposition, periods tend to have a proportionally larger sidebearing than other characters. Thus the visual result is a modified French spacing. In the United States, where many abbreviations are set with periods, this extra space in the side bearing may lead to rather loose composition. Linotron Bembo is particularly offensive in this regard. CME

Fresh recto. *See* **New recto; Recto.**

Frontispiece. An illustration printed or tipped in (pasted) facing the **title page.**

Front matter (FM). The pages in a book preceding the main text (also called *preliminaries* or *prelims*). These pages may include **bastard title, series title, title page, copyright page, dedication, epigraph, preface, acknowledgments, contents,** and others. They are usually numbered with lowercase **roman numerals.** This practice allows changes to be made in the paging of the front matter without disturbing the arabic pagination of the main text and thus is particularly useful when a page must be gained or lost. *See also* **Appendix 2.**

Full caps. Instruction to set copy in all **capital** letters of the same **font.**

Full measure. Usually the line length of the main **text page,** as in the instruction, "extracts set full measure." If the **type page** is wider than the text page, as when there are **sidebar** columns, this term can be confusing.

Full page. Depending on the context, *full page* may mean any of the following: (1) An instruction to leave a blank page for **artwork** that has not yet been provided; (2) an instruction to position sized art without text on the same page even if text would fit; (3) an instruction to size, shoot, and position art so that it fits within the **text page;** (4) the size of the text page.

G · · · · · · · · · · · · · · · · · · ·

Gallery. A section of **illustrations** grouped in a succession of pages rather than placed individually throughout the text. The gallery may or may not be counted in the book's page-numbering sequence; if it is counted, it is customary to use **blind folios.**

Galley. Often used as synonym for **galley proof.** (1) In photocomposition, the photographic paper used in the typesetting machine typically comes in 150-foot rolls. As the type is set, it is necessary to have breaks in the text stream so that the **repro** can be photocopied to make proof. One kind of break is the **page;** the other is the galley, which is just as long as can be conveniently photocopied to make galley proofs – usually about 13 inches. By convention, a *tagline* or *header* identifying the job is set at the top of each galley, along with the galley's number. (2) With metal type, the galley is a metal tray used to store type.

Galley proof. Proof of text before it is made up into **pages.** *See also* **Galley.**

Gang. An instruction to put two or more images that fall on the same page or **spread** together – both at the head of the page, if that is the style, as opposed to one at the head and one at the foot.

Generic code. A generally understood code put into an **electronic manuscript** by the publisher or author that mirrors the handwritten annotation on the **hard copy** or in the **specifications,** such as CT for the chapter title and A for the **A-heads.** To enable the type shop to use a **global search and replace** with the specific codes or key sequences and to avoid the possibility of codes' being mistaken for text, or vice versa, generic codes are usually marked by backslashes, brackets, etc.; for example, \CT, {CT}, or ⟨ct⟩.

Ghosting. *See* **Backing up.**

Global search and replace. Instruction to run a computer search for all instances of a particular character or string of characters and replace or change them without operator intervention or decision making. Just how sophisticated this operation can be depends both on the computer program used and on the imagination of the user.

Glossary. List of foreign words or technical terms, with translations or definitions, sometimes included in the **back matter** of a book. *See also* **Appendix 2.**

Golden section. A set of proportions based on the ratio 0.610:1, in which A is to B as B is to the sum of A plus B (see diagram). Thus a sheet of these proportions, folded in half over and over again, will maintain the relative proportions of width to depth. The golden section is fundamental to much traditional book design from the Renaissance onward and has been elaborated to include the relationship of page size, text block, and margins.

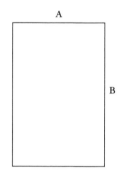

Lucid discussions of the golden section as it relates to book design are included in Bringhurst, Elements of Typographic Style, *and Tschichold,* The Form of the Book *(see Bibliography).* RE

Gothic. (1) The type style **black letter.** (2) Certain heavy nineteenth-century **sans-serif** faces are called gothics, perhaps because their color recalls black letter. The more recent typeface News Gothic refers to this tradition. (3) In text setting, a character used to suggest a shape (as in A-frame or V-formation) may be marked *gothic.* This conveys that it is to be set sans serif, in contrast with the text. *See also* **Appendix 3.**

Grave accent (`). *See* **Accent; Appendix 7.**

Greek aspirates. *See* **Breathing.**

Greeking. A technique in design using meaningless squiggles to simulate text in layouts.

Grid. A modular system used for overall design and **layout** planning in which the design area is divided into a pattern of consistent shapes. Some proponents of the grid system hold that all elements of all pages should relate exactly to the grid. Others believe that the grid may be useful to establish unity among the elements of a book but that it is possible to depart from it to resolve specific problems. Overdependence on the grid can produce rigidity or sterility.

Grid sheets. *See* **Boards.**

Grotesque. The name of certain typefaces without **serifs,** but usually with some slight variation in stroke weight. *See also* **Appendix 3.**

Guillemets («», ‹›). Form of **quotation marks** used in French, Italian, Spanish, Russian, German (where the order is transposed »as here«), and other languages. Also called *duck feet, French quotes.*

Gutter. (1) The **gutter margin.** (2) The space between columns of type, as in a table, an index, two-column text format, etc. *See also* **Ditch.**

Gutter margin. The inside margin of a page, closest to the binding and opposite the **outside margin;** also known as the *back margin. See also* **Margin.**

H

HA. House alteration. *See also* **Alteration.**

Haber rule. *See* **Type gauge.**

Haček (ˇ). Diacritic used in Czech. Also known by the names *caron* and *wedge* and used in Slovak, Croatian, Latvian, Lapp, etc. *See also* **Appendix 7.**

Hairline. In some **modern** typefaces, the very thin part of the letter. *See also* **Appendix 1.**

Hairline bar. *See* **Crossbar; Appendix 1.**

Hairline rule. A light, thin **rule,** usually ¼ point in thickness; also called *fine rule.*

Hair space. The thinnest character spacing material used in **handset type,** not a precise measurement. To a compositor, the instruction to add or delete a hair space basically means "improve the look by adding or subtracting a little space." *See also* **Thin space.**

Half title. The main title of a book, which usually begins arabic page 1 and is identical in wording to the **bastard title.** Also called the *second half title.* Sometimes dropped in order to save two pages. To allow for this possibility the half title can be numbered as the last page of the **front matter** rather than the first arabic page so that, if it is dropped, changing text folios can be avoided. *See also* **Appendix 2.**

Halftone. The translation of **continuous-tone art,** such as a photograph, into a printable image. The varying shades of gray are represented by a series of evenly spaced dots of different sizes and shapes. *See also* **FPO; Line art.**

Half uncial. Calligraphic letterforms that evolved from ancient Roman everyday documentary handwriting and developed toward the modern **lowercase** alphabet.

Hamza (’). Mark used for a glottal stop when setting transliterated Arabic or Hebrew.

H & J. Abbreviation for *hyphenation and justification,* one stage in the composition process. Running H & J "explodes" the codes (that is, instructs the computer to address a complete set of instructions) and generally sets the type according to **specifications** that have been programmed into the computer for a particular job. There is at present no H & J program that does not require knowledgeable checking to achieve high-quality composition. *See also* **Justify.**

Handset type. Individual **characters** of metal type assembled (composed) by hand into words in preparation for printing.

Handwork. Any work that cannot be routinely carried out by the typesetting program used; for example, setting genealogical tables or charts. What is considered handwork will vary from typesetter to typesetter.

Hang. A specification used when it is difficult to give an exact value. In the instruction to "hang the chapter title from the first text line," the intent is that the **ascenders** of the chapter title will align with the ascenders of text on facing or backing pages. *See also* **Flush and hang.**

Hanging figure. *See* **Old-style figure.**

Hanging indention, hang and indent. *See* **Flush and hang.**

Hanging punctuation. Punctuation marks positioned beyond the outer edges of the text measure or **paragraph indent**. The instruction may be to hang the punctuation only on the left (usually **quotation marks**) or on both the right and the left. If punctuation is to be hung on the right, it is necessary to specify which kinds of punctuation marks are to be hung.

Usually only the less prominent marks are hung on the right, such as the period, comma, and hyphen. Those marks that span the x-height or cap height, such as the colon and the question mark, are not hung, because they are substantial enough to maintain the visual alignment of the type block. RE

Hang quotes. Instruction to position **quotation marks** beyond the type measure so that quoted text material that begins or ends a line will not be indented within the text block but will align with the following lines. Quotation marks can be hung left only or left and right. They may also be hung within a **paragraph indent** so that, in works with mixed dialogue and text, the letters starting the paragraphs will align.

Hard copy. **Manuscript** printed out on paper that is a duplicate of an **electronic manuscript**; may also be referred to as a *printout* from a tape or disk. With some electronic manuscripts there is no hard copy.

Hard hyphen. A hyphen in a computer file that should be set whether or not the word falls at the end of a line, as in "mother-in-law." The hard hyphen is a different character from the **soft hyphen** in a computer file.

Hard return. In a computer file, a "carriage return" that should remain, such as an end-of-paragraph return, as opposed to a **soft return**, which should be removed if more copy will fit on the line when the file is edited or reformatted.

Hatchmark. *See* **Number sign; Pound symbol** (#).

Head, heading. Any of several elements treated differently typographically from **body copy,** such as **chapter heads, subheads, running heads,** and **column heads.**

Header. *See* **Galley.**

Head margin. The space from the top edge of the trimmed page to the top of the print area (usually the **ascenders** of the first line of type, whether it is a **running head** or text); also called *top margin.* Printers prefer this measurement to be given in inches, though it is often given in **picas** and **points.** The head margin may also be specified with reference to the **baseline** rather than the ascender height of the first type element. This is acceptable, as long as the intention is made clear, and may be the less ambiguous method when running heads on **facing pages** have different ascender heights – for example, if small caps are used on the verso and full caps and lowercase on the recto. *See also* **Margin.**

Head trim. *See* **Trim, trim size** (2).

Hickey. Dirt or blemish on **repro** copy, **bluelines,** or printed material.

High-resolution proof. Proof where the images (usually typographic characters) are extremely clear and sharp. *See also* **Dots per inch; Resolution.**

Hold in. Instruction to **indent,** as in "hold in extracts 1 pica left only."

Hook. *See* **Ogonek.**

Horizontal format. *See* **Oblong.**

Horizontal rule. *See* **Rule.**

Hot-metal composition. Typesetting using a machine that casts metal type from molten alloy as it is needed, such as any of the **linecasters** or a **Monotype** machine. Handsetting type with foundry type is not, properly speaking, hot-metal composition because foundry type is stored ready to be used and reused.

House alteration (HA). *See* **Alteration.**

House font. General purpose or "stock" typeface, used in shops that specialize in work where the client specifies no type style but leaves it to the shop to decide what type to use. The **fonts** will include a

serif roman such as Times, a **sans serif** such as Helvetica or Gill, and a **script** font.

House style. A publisher's or type shop's stylistic guide for **copyediting,** typesetting, or paging. Considered by compositors as a supplement to the designer's specifications for a particular job. *See also* **Style sheet; Appendix 5.**

Hungarian umlaut ("). Less frequently known as a *double acute* accent, though the latter is a better term, since the "regular" **umlaut** (¨) is also used in Hungarian. *See also* **Appendix 7.**

Hung punctuation. *See* **Hanging punctuation.**

Hyphen. *See* **Dash; En dash; Hyphenation.**

Hyphenation. In composition, breaking words at the end of a line according to rules. Most computer typesetting programs have a built-in hyphenation program. *See also* **Consecutive hyphens; Hard hyphen; Soft hyphen.**

Hyphenation and justification. *See* **H & J.**

Hyphenation dictionary. (1) Stylistic list showing the way words are to be hyphenated. (2) File used in computer composition for the "program" to look up the proper **hyphenation** of words. *See also* **Exception dictionary.**

Hyphen block. Proofreaders' mark indicating that the number of **consecutive hyphens** is greater than allowed. *See also* **Break block.**

I

Ictus. *See* **Scansion marks.**

Idiot file. In computer composition, a file containing all the text and typesetting codes, but one where **H & J** has yet to be run.

Ikarus system. One of several computer systems for originating digital typefaces.

Illumination. The art of hand-decorating initials, borders, or headpieces of book or manuscript pages using a variety of colors or gilding.

Illustration. A drawing, photograph, graph, map, etc., in the form of **line art** or **continuous-tone art,** used to clarify or decorate a text. Illustrations may be **figures** printed along with the text they refer to or may be assembled in a group. *See also* **Gallery; Plates.**

Imagesetter. A device for reproducing computer-created images (text or illustrations) on paper, photographic paper, film, or printing plates. The images are made up of a series of dots, and the quality of the final image is measured in **dots per inch** or **resolution.** Imagesetters are used today in place of line printers, typesetters, and, increasingly, platemakers. In addition to their ability to produce very high quality typeset material, most can also reproduce **tints** and screens, permitting **line art** and **halftones** to be set in place. An imagesetter is made up of two parts, the device that translates the image into dots (raster image processor or RIP) and the device that puts the dots on the paper, film, or plate (the marking engine). Examples of imagesetters include the Apple LaserWriter and the Linotron 300.

Imprint. The publisher's name, placed on the **title page.** The date of publication, address, and **logo** may also be included.

Indent. To begin or end a line or group of lines with a given amount of white space from the edge of the **text page,** such as the **paragraph indent** or the indention of an **extract** from the right and left margins; often used as shortened form of *indention. See also* **Flush and hang.**

Index. An alphabetical listing, at the end of a book, of names, places, and subjects included in the **text** along with the page reference where each entry can be found.

Inferior. *See* **Subscript.**

Infinity symbol (∞). A mathematical symbol, usually placed on the copyright page or with the **CIP** data to indicate that the paper used in the book is acid-free and meets American National Standards Institute (ANSI) standards.

Initial cap and lowercase. Specification to set the first letter of the first word as a **capital,** with the following text in **lowercase** (except proper names and other words that normally carry a cap). Distinguished from **caps and lowercase,** where the first letter of every principal word is set in caps, with the rest lowercase. These instructions are almost always used with headings or titles rather than with text.

Initial letter. *See* **Decorative initial; Drop cap; Raised initial.**

In line. Instruction to set something, such as a mathematical equation, in the text line rather than to **display** it.

Inline letter. Type ornamented with a continuous white line within the **main stroke.**

Insert. New copy that is inserted into the text at a proof stage.

Inside margin. *See* **Gutter margin; Margin.**

Intercharacter space. The normal space between letters as designed for a particular typeface. This may be changed by **kerning** specific letter combinations or by general **letterspacing** or **tracking.** *See also* **Kern.**

Interlinear. Referring to material situated or inserted between lines; for example, an interlinear translation, in which alternate lines are set in different languages, perhaps in different typefaces or sizes.

Interline space. *See* **Leading; Line spacing.**

International Phonetic Alphabet (IPA). An alphabet of several hundred characters used to provide a uniform system for transcribing the speech sounds of all languages.

International Standard Book Number (ISBN). Number assigned by a publisher (under a system established by R. R. Bowker and the International Standards Association) to a specific edition of a specific book. The ISBN usually appears on the **copyright page** as part of the **CIP** and on the back of the book's **dust jacket** or cover. It is used by publishers to keep computerized inventories and by bookstores, libraries, and wholesalers for ordering.

Interword space. *See* **Word space.**

Introduction. A part of a book that provides explanation, information, or comment preliminary to the main portion of the work. An introduction may be included in the **front matter,** paginated with roman numerals, or may be considered part of the main text and paginated with **arabic numerals.** Sometimes the introduction is, in fact, the first chapter. *See also* **Appendix 2.**

Inverted commas. *See* **Quotation marks.**

Inverted exclamation point (¡). Used in Spanish to precede an exclamation.

Inverted question mark (¿). Used in Spanish to precede a question.

IPA. *See* **International Phonetic Alphabet.**

ISBN. *See* **International Standard Book Number.**

Ital. Short form for **italic.** *See also* **Appendix 8.**

Italic. The style of letters in a type **family** that slope forward, as distinct from the **roman,** which are upright. Italic letters are also usually more delicate than roman letters, with less variation in stroke weight, and when the characters are not derived from **linecaster** designs, they are usually narrower than the roman. The first italic typefaces were made in Italy in the sixteenth century and were modeled on fifteenth-century handwritten forms. These typefaces were used to print entire books, small and easily transportable – in effect, the first pocket books. Thus the earliest italic fonts stood by themselves and were not considered part of a type family. *See also* **Cursive; Oblique; Type style.**

J ·

Jacket. *See* **Dust jacket.**

Jacket flap copy. Also called *blurb.* Copy that conveys information about the book, author, and publisher and becomes part of the design of the jacket. Flap copy is one of the few elements where it is not automatically assumed that the type is to be set **justified;** unless it is otherwise specified, many type shops will assume **ragged composition** because of the narrow measure.

Job font. Colloquially, a typeface used mainly in advertising, brochures, etc., and not generally considered appropriate for traditional bookwork.

Job type. *See* **Job font.**

Justify. An instruction to specify or set type so that both the left and right sides of the copy will be aligned. This is accomplished by evenly adjusting the spacing between words, characters, or both in a line – though any visible adjustment of the space between letters is not acceptable in high-quality work. With **handset type,** the even adjustment of space between words can be to some extent optical rather than mechanical; only a few **photocomposition** systems allow for such niceties. *See also* **H & J; Letterspacing; Word space.**

Keep standing. Instruction given when type matter is to be kept from one printing to another. Computer (phototypesetting) files are much easier to store than metal; even so, one should ask about the typesetter's practices if one wishes to store the files (type) for future use. *See also* **Standing type.**

Note that it is difficult to obtain insurance covering the data on a floppy disk. For this reason a shop may inform the customer that it will keep a copy of the job but cannot guarantee it will be available. Another common practice is to ask the type shop to send a copy of all files to the publisher, who can then assume responsibility for storing them. Even here the type shop will not always guarantee the usefulness of the files for setting type in the future. CME

Kern. (1) The part of a letter that extends beyond the **body** of the type, overlapping an adjacent **character;** common in **italic, script,** or **swash fonts.** (2) To adjust the space between individual characters. Unlike **letterspacing** or **tracking,** which are blanket specifications applying to all letters in a block of copy, kerning is a selective process used between character pairs. In computer composition, kerning tables or kerning programs are written so that the operator does not have to enter the kern by hand. Since either a positive or a negative value may be entered in these tables, *positive kern* is no longer the oxymoron it seemed in 1960. *See also* **Mortise.**

Kerning. The process or result of space adjustment between letters. *See also* **Kern; Track, tracking.**

Key. (1) Definition of symbols used on an illustration such as a **figure** or map; sometimes called **legend.** The symbols and accompanying explanation may appear either within the illustration or below it. (2) Codes or symbols used to identify separate pieces of copy on a **dummy, insert,** or **layout.** (3) The principal or master layout for a **grid** that is used for pasting up. (4) Short for **keyboard,** as in "the job has been keyed but not proofed."

Keyboard. To "type" or key in manuscript copy on a computer or terminal so that the words may be saved in a file for eventual use in computer composition.

Key in. *See* **Keyboard.**

Keyline. An outline, usually a thin line of red or black, positioned on a **layout** or **repro** to indicate the dimensions and exact position for an illustration or other element. Often a keyline is used as a substitute for a **window.** According to instructions, the keyline may or may not print. *See also* **Mechanical.**

Kill. Instruction to delete or destroy erroneous or unwanted material.

Knock out. Instruction that type or other copy is to **reverse** and appear as white on a darker background or in a dark area of a photograph.

Kroužek. *See* **Overring.**

Label. (1) Identification affixed to artwork, type, or other material indicating direction, size, percentage of reduction / enlargement, or other instructions. (2) A term sometimes used with copy to be set for a map or figure, as in "set map labels in 8/8 Helvetica."

Landscape. (1) British term for an **oblong** page or format, in contrast to **portrait**. (2) A desktop-composition term for a horizontal rather than a vertical page. *See also* **Broadside.**

Latin alphabet. The alphabet used in modern English and Western European languages, adapted from the Greek.

Lay of the case. The plan or scheme of arrangement of the type **case** used for handsetting metal type, in which each letter or character has a specific and unvarying location.

Layout. A hand-drawn or computer-generated drawing or sketch of a proposed design showing the position and appearance of all elements and used as a working diagram or shown for approval. Also called *mise-en-page. See also* **Mactissue.**

Lay out. To arrange material in the order in which it will appear in final form, as in making pages from galleys.

lc. Abbreviation for **lowercase.**

LC. Abbreviation for Library of Congress.

LC number. Identifying number of a book, assigned by the Library of Congress and included as part of the **CIP.**

Lead. *See* **Leading.**

Leader. A series of periods or short dashes, evenly spaced, used to guide the eye across the page from one element to another. Now generally considered unnecessary, these formerly were widely used on contents pages to connect chapter titles with their page numbers. Leaders are set from right to left and always align vertically, so deleting or adding a period affects the left end of the leader.

Leading. (1) In photocomposition, the amount of vertical space between lines of type, measured from **baseline** to baseline. In the United States and the United Kingdom the measurement is given in Anglo-American **points.** Most of Europe uses **Didot points,** with a few countries such as Germany and Switzerland using the **metric system.** In photocomposition, leading is synonymous with **linefeed,** *baseline skip,* or *interline space;* for example, 10/12 indicates 10-point type with a 12-point linefeed. The term has a different usage in metal composition, where leading is the extra space added to the **body size.** (2) With handset metal type, the thin strip of nonprinting brass, lead, or other material used to create additional white space between lines of type.

Leading scale. *See* **Line gauge.**

Leaf. Each separate sheet in a book, including a **recto** backed up by a **verso.** *See also* **Page.**

Legend. (1) Descriptive text identifying or explaining a **figure** or **illustration** and usually appearing directly below it; often used interchangeably with **caption.** (2) On a map or chart, the **key** to the elements used, placed either within the art or below it.

Letterfit. *See* **Fit.**

Letterspacing. Adding or, less frequently, subtracting space between the letters of a word. Usually, the **word space** value must also be adjusted. Letterspacing is usually specified by the designer, especially when words are to be set in **full** or **small caps.** Letterspacing values may be given in **points** or in **units.** When units are used, the base system (18 units to the em, 54 units to the em, etc.) should also be noted for the typesetter's reference. The instruction to "match layouts" is also acceptable. What causes a problem for the typesetter is when the written **specification** does not match the **layout,** and in this case the primary specification should be indicated. *See also* **Minus letterspacing; Track, tracking.**

Some compositors will slightly letterspace full and small capitals even when this is not explicitly specified. A traditional composition nicety is to set combinations like RY without kerning or letterspacing and to space other combinations, such as TI or IN, so that they have about the same optical space as the RY combination. In effect this amounts to letterspacing capitals. If a designer prefers that TI and IN be set using their natural spacing and RY be kerned to achieve optical balance, the specification "set close" should be used. Letterspacing is also sometimes used by typesetters as a help in justifying a line, though this practice is frowned on. Letterspacing of ⅟25 of a point to justify a line probably will not be apparent; any more probably will be and should not be used. CME

Library of Congress (LC). *See* **CIP; LC number.**

Ligature. Two or more letters combined into a single **sort,** or **character.** Often called **tied letters.** Originally the term was used to denote only the stroke that ties two characters together, such as the crossbar between the f and i in fi. *See also* **Diphthong; Appendix 1; Appendix 7.**

Lightface. A weight of letter lighter than that normally used for continuous reading. Usually a typeface intended for use in display rather than text.

Line art. A black-and-white **illustration** that has no gray values, such as a pen-and-ink drawing. Unlike a **halftone,** line art is usually pasted directly onto the **repro** or **scanned** in with the type. *See also* **Continuous-tone art.**

Linecaster. Generic name for typesetting machines such as the **Linotype** and Intertype machines, which cast a line of type with a single injection of molten metal.

Linefeed. The most accurate, but not the most common, term for specifying the space between lines in computer composition. The more commonly used term **leading** is acceptable as long as one remembers it should be used as a **base-to-base** measurement with computer composition. Modern computer composition allows for $\frac{1}{10}$-point increments in linefeed.

Line for line. Specification indicating that copy should be set in separate lines exactly as typed or marked, regardless of other text measure specifications. Verse **extracts,** for example, are almost always set line for line. *See also* **Alignment.**

Even such an apparently simple instruction can raise questions, for example, when there are turned lines of verse in the typescript. The typesetter does not know whether these turned lines should be run in if there is room on the typeset line or whether the manuscript turns should be preserved. In the absence of an instruction to run in, copy marked "line for line" will have such turned lines preserved in typesetting. CME

Line gauge. A ruler or an acetate sheet calibrated in various **linefeed** or **leading** increments. Also called a *leading scale.* Typical line gauges will show from about 6-point to about 20-point leading.

Line length. The length of a line, measured in a typescript by the number of **characters** in a typical line; in typeset material, measured in **picas.**

Line spacing. *See* Leading; Linefeed.

Lines per inch. *See* Dots per inch; Halftone.

Line up. Synonymous with *align* or **align on,** but usually used as a proofreader's mark rather than a designer's instruction.

Lining figure. Also called *aligning figure, capital figure, modern figure, ranging figure.* Lining figures are numerals all of the same height, usually the same as that of the capital letters in a typeface, though in some fonts they are slightly shorter. Lining figures are generally used for tabular matter, for **superscripts** and **footnote calls,** with capital letters (in display or in constructions such as "Interstate I-40"), and with mathematical expressions even when **old-style figures** are specified for the job. If old-style figures are intended in any of these cases, an instruction is needed.

In a job with a significant number of mathematical expressions, the instruction "use lining figures with math" leaves in doubt which style to use with simple expressions such as "2%" or "2 percent." CME

Unless directed otherwise, typesetters will generally use lining figures, especially in setting tables. Many designers feel that tables are better set in old-style figures. RH

Linotype. A **hot-metal** typesetting machine that combines the casting of type with its composition, as opposed to the **Monotype** machine, in which the two operations are sequential. Pressing a key on the machine drops a brass **matrix** (mold) in place on the line being set. The **spaceband** key drops expandable wedges, with which the operator **justifies** the line when it is nearly full. Molten lead is then forced into the molds, and the entire line is cast as one piece of metal or slug (hence the generic name **linecaster**). The cast line is then placed on the **galley** tray, and the individual matrices are automatically distributed back to their places. Linecasters such as the Linotype machine allow type to be set more quickly than the Monotype machine, but there are some inherent compromises in quality that are not present in either **handset type** or Monotype composition.

Live matter. Material that has not yet been used, is in process, or should be saved for future use. *See also* **Dead copy, dead matter.**

Locking up. The process by which type, illustration blocks, or other material is fixed tightly into a rectangular metal frame (a chase) with wedges (quoins) and strips of wood (furniture) in preparation for letterpress printing.

Logo. Short form for **logotype.** Now used with particular reference to marks of corporate identity or a word or combination of initials treated as an entity, such as IBM or CBS. The term is also commonly used to refer to an identifying symbol, either an abstract graphic or a pictorial device.

Logotype. Letters cast from a common **matrix,** such as To or Te, to achieve the effect of a **kern.** Most common with **Linotype** machines, where it is not possible to have a kern as a property of a **character,** since the matrix cases butt on the line. *See also* **Ligature.**

Long and/or short. A specification that allows a deviation from normal column or page length when pages are made up. Flexibility in making up pages is necessary to avoid **widows, orphans,** or other **bad breaks.** When **facing pages** are to align, as opposed to having ragged bottoms, and when **carding** is prohibited, allowing the page depth to vary is the obvious choice. Pages – preferably facing pages – may be allowed to run either long or short, but there is a firm convention that a short **spread** should not be followed by a long one, or vice versa. *See also* **House style; Page makeup.**

Long descender. The portion of a **lowercase character** that falls below the **baseline,** designed to be especially long. Most **foundry type** originally designed for bookwork had long **descenders,** since these would give adequate space between the lines of type without the need to insert extra space by hand. With the advent of other printed material, like newspapers, it was common to select a **typeface** with **short descenders,** such as Plantin, which would allow more copy to be set in a given space. A refinement in typefounding was to offer some **fonts** with both short and long descenders, a practice that has carried through today for some **photocomposition** typefaces, such as Linotron 202 Caledonia.

Long page. A page longer than the specified page length. In the past, when the type shop had to run a long or short page it was marked on the proof, but today this is not always done. Since long or short pages must be taken into account by the editor or designer if re-paging is needed, it is highly desirable to have the type shop mark them. Both the designer and the proofreader look at proof page by page, and it is easy to miss long or short pages.

Long primer. A term used before the general adoption of the **point** system for a type size nearly equal to 10-point.

Longum (¯). A **scansion mark** used to indicate a long syllable in prosody. Distinguished from a **macron** (ˉ), which marks a long vowel. A longum is wider than a macron and is usually placed above the **cap height.** In transliterated Greek prosody, both a macron and a longum may be used. *See also* **Appendix 7.**

Loop. A rounded form in a letter, as in the Baskerville lowercase roman g, that is not closed and is less circular than the **bowl** (the upper part of the g). *See also* **Appendix 1.**

Loose blues. *See* **Blueline, blue.**

Lowercase (lc). The small letters in type as distinguished from **capitals** or **small caps.** *See also* **Case.**

Low-resolution proof. *See* **High-resolution proof.**

Lozenge. An **ornament** having four equal sides, two obtuse angles, and two acute angles. Thus there is no single shape for the lozenge (one example, ♦). *See also* **Diamond.**

ℳ ·· · · · · · · · · ·

M. *See* **Em.**

Machine condensed (expanded). Referring to the electronic manipulation of type to make it narrower or wider than the original design. The vertical strokes are thinned or thickened while the horizontal strokes remain the same as in the original character. This distortion is not present in designed condensed (expanded) **true-cut** fonts.

Machine italic. The creation of an **italic** by electronically sloping or slanting the **roman** letters. Although this technique is sometimes effective for design, it is not considered proper for a typesetter to "slope the roman" as an alternative to purchasing the italic typeface. *See also* **Computer generated; Oblique.**

One special situation is the machine generation of italic small capitals, which are almost never cut in photocomposition fonts. One way to produce italic small caps is to slope the roman true-cut small caps, resulting in a letterform that is too heavy as well as looking like a sloped roman rather than an italic. Unfortunately, the other possibility is to start with italic full caps and reduce their size, resulting in letterforms that are too thin and distorted. A potential solution for PostScript fonts is given in the comment under the entry fake small caps. CME

Machine small caps. *See* **Fake small caps; Machine italic.**

Machine space. Mechanically defined and fixed space of various increments. *See also* **Visual space.**

Machine unit. The smallest unit of **escapement** or movement on a typesetter or **imagesetter,** which determines both the **resolution** and the degree of character fitting possible on a given machine. For example, the Linotron 202 has a machine unit of $\frac{2}{27}$ of a point. *See also* **Fit.**

Machine zero. *See* **Monoline zero.**

Macro. A routine or small computer program that can be called by very few keystrokes. Macros allow the type shop to provide custom typographic refinements without the cost of doing the work by hand.

Macron (ˉ). A horizontal line over a vowel indicating that it is long. *See also* **Longum; Appendix 7.**

Mactissue. Layout or **sample page** generated on a computer rather than in the traditional way using a drawing table, T-square, drawing tools, and tracing (tissue) paper. Computer-generated layouts cannot always be exactly replicated by a typesetter who might be using different equipment and fonts. Mactissues should be viewed as an alternative to tracing and are to be taken as a general guide only, unless specifically stated otherwise.

Main stroke. The most prominent part of a **character**; also called **stem.** *See also* **Appendix 1.**

Majuscule. A **capital** letter or **uncial.**

Makeup. *See* **Page makeup.**

Manuscript (MS, plural MSS). Literally "written by hand," but used for the author's original copy, whether typewritten (typescript) or a computer printout (**hard copy**).

Margin. The white space between the printed matter and the edge of a page. There are four margins: the **head** or *top* **margin;** the **gutter,** *inside,* or *back* **margin;** the **outside,** *face, fore-edge,* or *thumb* **margin;** and the **bottom,** *tail,* or *foot* **margin.** Usually only the head and gutter margin are specified. In asymmetric typography, however, the left and right margins may be specified, causing the gutter and outside margins to vary on **verso** and **recto** pages.

Marginal note. (1) Copy in the margin area of a manuscript page that, when circled, is not to be set; usually either author-editor dialogue or a note to the typesetter. It is important that the typesetter be able to tell quickly whether a marginal note is copy to be set. (2) Note to be typeset, frequently designed to be set in a different size or typeface from that of the main text. Usually placed in the **outside margin,** since the **gutter margin** is generally too small. Page makeup with marginal notes is usually more expensive than straight text only, and sometimes more expensive than text with **footnotes,** since the placement of marginal notes is usually not consistent.

Marginal sidehead. A heading or **subhead** in the marginal area of a page.

Markup. *See* **Type markup.**

Mary. *See* **Molly.**

Master. *See* **Cut.**

Master grid. The design and positioning framework for a project, set up by a designer, compositor, or printer. The master **grid,** which can be a tissue or acetate **layout,** a computer format, etc., should show **trim, margins,** column structure, and all other pertinent design information.

Master proof. A set of **galley** or **page proof** on which all the **printer's errors** have been marked and queries answered and all the author's, designer's, and editor's **alterations** have been added. These proofs are returned to the compositor for correction and revision and are considered the primary reference source for the work until superseded by further proofs, which in turn generate their own "master proof."

Match. Short form for "match an element of the layout." An instruction to the typesetter to match the size, style, letterspacing, etc. of a sample that is included on the **layout** or attached to the manuscript.

Matrix (plural *matrices*). (1) In mathematical composition, a rectangular table or array of numbers in which an isolated location can be identified by coordinates, or identifiable points of reference. (2) In **hot-metal composition,** the type mold used to cast type.

Measure. *See* **Type measure.**

Mechanical. **Camera-ready copy** of type, art, or both, with exact placement of all elements except **halftones,** which are usually indicated on the mechanical by **corner marks, keylines, windows,** or **FPOs.** Mechanicals for the interior of a book should include general instructions for positioning all pages, with corner marks on display pages. Mechanicals for a **dust jacket** usually include instructions to the printer indicated on an **overlay** of tracing paper. If it is necessary to write instructions directly on the artwork, a **nonrepro blue** pen or pencil must be used. *See also* **Boards; Layout; Register marks.**

Media conversion. Converting computer files between systems of incompatible media, for example, paper tape and floppy disk. Today these incompatibles have been greatly reduced, and the term is now frequently used to mean the same as **file conversion.**

Medium-resolution proof. *See* **High-resolution proof.**

Metal composition. The process of setting metal type, either by machine or by hand. With **foundry type** the individual **characters** (**sorts**) are supplied by the foundry, and the metal alloy is very hard. The job is composed by hand, letter by letter, and after the job (or a portion of it) has been printed, the type is cleaned and distributed to be used again – a lengthy process. With **hot-metal composition,** the foundry supplies the type molds or **matrices** to the type shop, and machines are used to cast the type as it is needed. Machine-cast type usually uses a softer alloy than is used in foundry type, limiting the number of copies that can be printed from it. Type from **linecasters** is usually cast softer than **Monotype,** and it was a common practice to make plates from linecaster composition if more than 4,000 or 5,000 copies of a work were to be printed. After printing, the type is melted down and the metal alloy reused. *See also* **Linotype.**

Metric system. A decimal system of measurement developed in France in 1790 and used particularly in Germany and Switzerland for typographic specifications.

Mid. *See* **Middle space.**

Middle English. The English language from about 1100 to 1500. *See also* **Old English** (2).

Middle space (mid). In hand composition, a 4-to-em space. Thus, in a line of 12-point type, a mid space would equal 3 points or approximately 4.5 units in an 18-unit system. *See also* **Unit.**

Minuscule. A **lowercase** letter.

Minus leading. Using a **linefeed** value (**leading**) less than the specified point size of the type, for example, 8/7. This is possible only with computer typesetting. The specification is most often used for full **capitals** or **small capitals,** or for **display** lines where the particular combination of the letters allows for such adjustment. It is readable in lowercase settings only with fonts where the ascender-to-descender measurement is significantly less than the nominal point size. For example, Linotron Garamond no. 3 (12-point master) measures 15.5 points **ascender** to **descender** with a nominal 18-point size specification. If desired, it could be set 18/17 and still have 1.5 points clearance between the lines.

Minus letterspacing. The reduction of the normal space between **characters.** Also known as *white-space reduction* (*WSR*) or *negative letterspacing.* This is possible only with computer composition, where the most common use is for tightening the letterfit in large sizes set from small type masters. For example, the **fit** of the characters of many fonts between 16- and 36-point will benefit from using minus ⅟54-em **letterspacing,** and when set in 36- to 72-point, using minus ²⁄54-em letterspacing. This kind of decision is best left to the compositor, but if it is specified, **units** should be used. *See also* **Track, tracking.**

Mise-en-page. *See* **Layout.**

Mixing. The combination of more than one kind of type, more than one size, or more than one component of a type **family.**

Mnemonic coding. In data storage and retrieval, an easily identifiable or remembered abbreviation for a format, instruction, or routine. For example, a line space (blank line) could be coded as \LN{1} and two line spaces as \LN{2}. *See also* **Generic code.**

Modem. Acronym for *MOdulate/DEModulate.* A device that permits information to be transmitted over telephone lines from one computer system to another.

Modern. Family of **typefaces** with contrasting thick and thin strokes and hairline **serifs.** *See also* **Appendix 3.**

Modern figure. *See* Lining figure.

Module. A standard pattern or unit of measurement used in standardized designs. *See also* **Grid.**

Molly. Antiquated oral term used in the noisy old print shop – along with *mary* and *mutton* – to distinguish an **em** from an **en** (called **nut**).

Monogram. A **sort** composed of one decorative letter or several letters, often interwoven or intertwined, used as the abbreviation of a name. *See also* **Logo.**

Monoline. A typeface or character designed with uniform thickness (or nearly so); for example, Futura. *See also* **Sans serif.**

Monoline zero (o). Numeral of uniform weight (also called *machine zero*). Some typefaces use a monoline character for the **old-style figure** to distinguish it from the lowercase o of the alphabet. For example, the Monotype Sabon old-style zero is monoline (o), and some designers and typographers prefer to substitute the lowercase o for this monoline zero.

If a designer wants to substitute an o for a zero, this should be clearly requested in the design specifications. Some type shops will be unable to make the substitution; others can make it only before coding, or only before certain codes are "exploded," since zeros may well appear in coding schemes, and these must stay as zeros. Finally, unless the o just happens to have the same set width as the numbers, the monoline zero must be used in tables to avoid losing decimal alignment. Given these problems, one reasonable compromise might be to substitute the o for a monoline zero in display but retain it in text. Where PostScript fonts are being used, some type shops can prepare a custom font to resolve these problems. CME

Monospacing. Setting with an unvarying amount of space between elements, usually when all letters have the same **set width,** as with a typewriter.

Monotype. A typesetting machine consisting of a keyboard whose operation produces perforations on a roll of paper and a caster that casts and assembles individual pieces of type and spaces in the order determined by the perforations. The term can also be used for the work produced from a Monotype machine or the printing done from that work. Monotype composition involves far fewer typographic compromises than **Linotype** composition, since it is possible to **kern** letters, and **italic fonts** may have **set widths** different from those of **roman** fonts. Since letters are cast individually, some **handwork** can be done. The primary disadvantage of the Monotype machine, compared with the Linotype, is that it is slower, making composition more costly.

Mortise. (1) In **handset type,** to "pull" letters closer together by cutting away metal. The term has been to some extent carried over into computer composition, in which **kern** is sometimes applied to either adding or reducing space between character pairs; mortise always means reducing space. (2) When **repro** paper is mounted on **boards,** to cut a hole in the repro paper so that another piece of repro paper with a correction fits snugly into it. It is wise to mortise corrections in repro paper when they are quite small (a letter, a word, etc.), so that they will stay in place. Also called *cutting in.*

MS (plural *MSS*). Abbreviation for **manuscript.**

Mutton. *See* **Molly.**

N. *See* **En.**

Nasal hook. *See* **Ogonek.**

Negative letterspacing. *See* **Minus letterspacing.**

New recto. Instruction to begin copy on a new right-hand page; also called *fresh recto.*

Nick. (1) In **photocomposition** systems in which the image is produced by a lens system, an extra hairline **serif** on the corner of a character that disappears on exposure, leaving a sharp, square corner. (2) In **foundry type,** a notch, usually on the underside of a letter, to help the compositor set the type right side up and detect any **wrong fonts.**

Nonbreaking space. A **word space** normal in every way except that it cannot be converted into an end-of-line break. For example, the space between a number and an abbreviation, such as "1 m" for "one meter," where a line break between the 1 and the m is to be avoided.

Non-Latin alphabet. Any alphabet other than the Roman or **Latin alphabet;** for example, Arabic, **Cyrillic,** or Hebrew.

Nonpareil. Archaic term for a size of type equal to 6-point (pronounced "nonprel").

Nonranging figure. *See* **Old-style figure.**

Nonrepro blue. A pale shade of blue that will not reproduce photographically, used for writing instructions on **artwork** or **repro.** *See also* **Boards.**

Normal aspect. *See* **Portrait; Upright** (1).

Note. *See* **Endnote; Footnote.**

Number sign (#). Typographic symbol used for *number* or *pound* or to refer to an unnumbered footnote. Also used in marking manuscripts or proofs to indicate space to be inserted. Also called *hatchmark. See also* **Reference mark.**

Numeral. A number expressed in either arabic or roman form. *See also* **Lining figure; Old-style figure.**

Nut. Antiquated typesetting term for an **en.** *See also* **Molly.**

O

Oblique. (1) Usually refers to a **roman typeface** that has been slanted or sloped to make an "**italic**" form. Not all oblique **fonts** are "mechanically sloped" roman fonts; some are designed (Trump, Joanna) and have subtle differences from the roman typeface. Examples can be found in almost all categories of type styles; for example, Electra (though the oblique font was soon replaced with a true italic – even though it is called Electra Cursive), Memphis (an **Egyptian**), Helvetica (a **grotesque**), and Futura (a **sans serif**). One tipoff to obliqued designs is that the lowercase a is frequently different from normal designs, in which this letter is two-tiered (a) in the roman and one-tiered (*a*) in the italic. Perhaps because of the possibility of machine generating an ersatz oblique font with **photocomposition** systems, many foundries and type books use the term "italic" for oblique fonts that have been drawn separately from the roman version. (2) To machine generate an italic font because the type shop does not have it, or because a designer is using a typeface for which no italic version has been cut. *See also* **Appendix 3.**

Obliqued roman type. *See* **Oblique.**

Oblong. With reference to a book or page, describes a format where the width is greater than the height; also called **broadside,** *horizontal,* **landscape,** or *album format.*

OCR. Abbreviation for *optical character recognition,* a computerized system that converts graphic symbols to electronic signals by "reading" character shapes from a typescript or printed document and creating a corresponding electronic file that can be edited and styled. *See also* **Scanner.**

Odd sort. *See* **Pi character.**

Ogonek (˛). Diacritic used with vowels in Polish, Old Norse, Navajo, and other languages. Also called *Polish hook* or *nasal hook.* Not to be confused with **cedilla.** *See also* **Appendix 7.**

Old English. (1) A term used in the United States (and to some extent the United Kingdom) for **black letter,** a group of typefaces that evolved from the broad-nib pen style of **gothic** lettering used by scribes in northern Europe; *see also* **Appendix 3.** (2) The English language from about 450 to 1100, also called Anglo-Saxon. Old English and Middle English (ca. 1100 to 1500) require a few **special sorts,** including cap and lowercase **edh** (Ð, ð), **thorn** (Þ, þ), **yogh** (Ȝ, ȝ), **wyn** (Ƿ, ƿ), **ash** (Æ, æ), and Old English **ampersand** (⁊ or ⁊).

Frequently the characters are more ornate than shown here, but the general shape is the same. As a last resort, an **old-style figure** 3 can be used for a yogh and an old-style or small-cap-height figure 7 for an ampersand.

Old face. *See* Appendix 3.

Old-style figure. A numeral related to the lowercase alphabet, with 0, 1, and 2 being of **x-height**; 3, 4, 5, 7, and 9 being of x-height plus **descenders**; and 6 and 8 going to **ascender** height. These numbers are best used with **lowercase** (and **small cap**) settings to keep the even "texture" of the line. Also known as *nonranging figures* and *hanging figures.* Designer Paul Rand has suggested the names "lowercase figures" for old-style figures and "capital figures" for lining figures. *See also* **Lining figure** for an explanation of when lining figures are used in a job for which old-style figures have been specified.

One-up proof. A proof that shows pages singly, not as a **spread.** *See also* **Two-up proof.**

Open letter. A letter in which only the outlines are drawn and interior spaces are partially or completely open. Also called *outline letter.* One example of a beautiful open letter design is Lutetia Open, designed by Jan Van Krimpen.

Open spacing. (1) Letter, word, or line spacing that is wider than usual. (2) The normal character spacing of a font such as Gill Sans, which sets rather "generously."

Operator. *See* Typesetter.

Optical center. (1) The position on a page, approximately two-thirds up from the bottom, that is considered the aesthetic point of balance. (2) The visual center of a line or block of copy, as opposed to its mechanical center, referred to in the inexact and subjective specification to "optically center" verse **extracts** or other blocks of type with lines of varying lengths. *See also* **Visual space.**

Optical character recognition. *See* OCR; Scanner.

Optical letterspacing. The visual (rather than mechanical) spacing of letters to achieve an even **color** or texture, usually in **display** type.

For example, an HI combination will require more space between the letters than an LY combination in the same word to make the **letterspacing** visually even. Like **optical center,** perfect optical spacing is a very subjective matter. *See also* **Visual space.**

Optical space. *See* **Optical center; Optical letterspacing; Visual space.**

Original. The source copy for **text, illustration,** or any other element.

Ornament. A single nonalphanumeric typographic element used for decoration or emphasis; also called *dingbat. See also* **Flower.**

Orphan. A very short line at the bottom of a page, or a word or part of a word on a separate line at the end of a paragraph. An orphan at the bottom of a page is generally considered less offensive than a **widow** (which occurs at the top of a page). When an orphan appears at the end of a paragraph, the length of the orphan should be slightly greater than the **paragraph indent.**

Older styles of composition – which prohibited orphans, widows, and hyphenated words breaking overleaf – usually permitted some rewriting of text by the compositor, such as changing the placement of an adjective or adding or subtracting an article. Since this practice is no longer acceptable, an orphan or hyphenation overleaf is considered preferable. CME

Like widows, orphans can often be avoided by asking the editor or author to rewrite to add or save a line. NBP

Outline letter. *See* **Open letter.**

Output. (1) To produce **hard copy** from a computer file. (2) The hard copy itself or, occasionally, the final result produced by computer composition in either paper, film, or magnetic form. Paper output is **camera-ready copy,** negative output is ready for the platemaker, and magnetic "output" will be "run out" by the printer as negatives or directly to printing plate.

Output resolution. Refers to how sharply a machine or system can render an image. *See also* **Density; Dots per inch; Imagesetter; Resolution.**

The output resolution currently advertised with imagesetters and laser printers invariably refers to only one part of the system. Statements like "the greater the resolution of the output device, the better the definition of the image" assume that all other factors are equal. Thus a Linotron 202 has a resolution of about 960 lines per inch, and some plain-paper laser printers have a resolution of 1,200 dots per inch. But the media are quite different, since photosensitive paper is capable of much higher resolution than powdered ink (toner) on paper. Note too that the highest resolution does not necessarily produce the most usable image. On the Linotron 202, overexposing and overdeveloping certain fonts in text-size settings – for example, Linotron Perpetua in a 10-point setting – will produce a printable image, whereas a "perfect rendering" of them may not. Overexposing and overdeveloping have the effect of fattening the letters (reducing the resolution), so that the thick parts of the letters as well as the thin parts will gain size and the once-sharp lines of the serifs will round. Nonetheless, in a 10-point setting the practice will produce a more usable image. CME

A similar effect can be achieved with imagesetters. Many imagesetter fonts have been designed to perfectly match metal type fonts without compensating for the slight spread of the ink when the characters were printed on letterpress. At very high resolutions these fonts appear too thin when printed by lithographic presses. Outputting at a lower resolution (about 1,260 dpi) helps these fonts regain some of their original character. RA

My original comment also applies to imagesetters. For example, when Monotype Bembo was released in PostScript form, it appeared too thin at small sizes. We experimented with different output resolutions – 1,200 dpi, 900 dpi, and 600 dpi – on a Linotron 200 imagesetter and with different exposures at each resolution. We found that increasing the exposure at 1,200 dpi produced a more usable image than simply lowering the resolution setting. This comment is added not to contradict the one by RA *above, but to show that a variety of factors are involved.* CME

Outside margin. The margin on the open side of a book, opposite from the binding; also called *face margin, fore-edge margin,* or *thumb margin. See also* **Margin.**

Overdot (˙). Diacritic used in Polish, Turkish, romanized Sanskrit, some Native American, and other languages. *See also* **Appendix 7.**

Overlay. A clear acetate or translucent vellum sheet, attached to **artwork** and keyed to it by **register marks.** An overlay may be part of the **camera-ready copy,** carrying elements of the type or design that are to print in a different color from the base artwork; it may also carry elements to be stripped together by the platemaker. A piece of artwork may require several overlays, depending on its complexity. An overlay is also used to carry instructions to the printer, such as the **trim size,** disposition of colors, and elements to be **reversed** out.

Overring (°). Diacritic used in Scandinavian and some Eastern European languages; also called *kroužek. See also* **Appendix 7.**

Overset, overset line. Type that exceeds the type measure specified. Occasionally, a designer will specify "OK to overset" when this is preferable to adding another line in order to stay within the measure.

P

Page (p.). One side of a **leaf** in a book. *See also* **Page makeup; Recto; Verso.**

Page break. *See* **Bad break.**

Page depth. *See* **Depth of page.**

Page description language (PDL). A computer program that acts as an interface between a computer graphics program (PageMaker, Quark Express) or typesetting program (Magna, TEX) and a laser **imagesetter.** Examples of PDLs are Adobe PostScript and HPL-III. The unique characteristics of PDLs are that they permit the simultaneous setting of text and graphics in place; they require the setting of type in full-page units with all elements in their proper position; the selection of fonts is not limited to those supplied by the output-device manufacturer; and they permit a wide variety of input computers and programs to interface to an even wider variety of output imagesetters.

Page makeup. (1) The arrangement and manual **pasteup** of type, illustrations, and any other elements, or their automatic machine **pagination** into final form. (2) The result of making up pages, as in "the page makeup was good (bad, acceptable)." *See also* **Bad break.**

Page number. *See* **Folio.**

Page proof. The final form of proof made up into pages, used for proofreading, checking corrections, positioning illustrations, etc. Page proof must include either **trim marks** or top-of-type-page **corner marks** where appropriate, so that the **sink** can be checked. Most book printers prefer guides at the top of the **type page,** but a few prefer trim marks.

Pagination. (1) The numbering of pages in a book, including both **roman numerals** for **front matter** and **arabic numerals** for the main text and **back matter.** (2) Making up pages (*see also* **Page makeup**).

Paging dummy. *See* **Dummy.**

Paragraph indent. An indention preceding the first word of a paragraph. Common paragraph indents are an em, a pica, and 1½ ems, but other values may be used.

Paragraph mark (¶). Also known as a *pilcrow.* In manuscript or proof, used to mark a paragraph break. The symbol is also used to indicate a numbered paragraph in citations of legal work, such as §10, ¶5. The pilcrow can also be used as a typeset device to mark a new paragraph when text is to be set with run-on paragraphs; usually an ornamental rendering is used, such as ¶ or ¶.

Parallels (‖). Double vertical lines used to indicate a reference to an unnumbered footnote. *See also* **Reference mark.**

Parens. Short form for **parentheses.**

Parentheses ( ). Upright curved strokes used in pairs to enclose matter incidental to the meaning of a sentence, asides by the author, or references within the text. *See also* **Brackets.**

Part title, part half title. Usually a **recto** page, with a **blind folio,** containing the number, title, or both of a part or section of a book and usually followed by a blank page.

Pasteup. The manual assembling of elements (type, illustrations, tables, etc.) into **camera-ready copy.** The term is usually used for the interior of a book, for **dust jackets,** and for cover art. *See also* **Mechanical.**

Patch. *See* **Strip-in correction.**

PDL. *See* **Page description language.**

PE. *See* **Printer's error.**

Penalty copy. Copy supplied for typesetting that is hard to read because it is dirty, very small, an out-of-focus photocopy, single-spaced, badly marked, on over- or undersized sheets, double sided, in a foreign language, or with many handwritten corrections. Such manuscripts require extra time to decipher and set and thus incur an extra charge by the typesetter. Charges can vary substantially, and suspect copy should be submitted to the supplier for assessment of penalty charges. *See also* **Clean** (1).

Phonetic alphabet. *See* **International Phonetic Alphabet.**

Phonetic symbols. Symbols that represent the sounds and other features of speech. *See also* **Appendix 7.**

Photocomposition. A photographic system of setting type. The evolution of photocomposition is often described in terms of generations. First-generation photocomposition machines were basically adapted **hot-metal** machines, fitted with a film master strip or other carrier of character images and a lens-projection system that allowed light to be flashed through the image to expose paper or film. Second-generation machines used rotating disks or film strips and a lens system to expose photosensitive paper. Third-generation photocomposition machines, such as the Linotron 202, form the character image onto a cathode-ray tube, which transfers it to photosensitive material by means of fiber optics or a lens.

Fourth-generation machines are differentiated by laser imaging and by the facts that the output (or **imagesetter**) machine can set either type or graphics and that the type may come from a variety of vendors. Linotronic imagesetters, for example, will set type purchased from **Monotype.**

Photostat, stat. A form of photographic copy made from a paper negative rather than a film negative. *See also* **PMT; Velox.**

Photo Typositor. *See* Typositor.

Pica. A standard typographic unit of measure equal to 12 Anglo-American **points.** Six picas equal 0.9936 inch, though personal computer systems round this off to an exact inch. This system of measurement is used primarily by typesetters, but also by many printers and binders. Historically pica was a **type size,** roughly equal to a 12-point setting. *See also* **Pica type.**

Pica pole, pica rule, pica stick. *See* Type gauge.

Pica type. The larger of the two common U.S. monospace typewriter faces, which yields 10 **characters** to the inch. *See also* **Elite type.**

Pi character. A **sort** or **character** other than **alphanumeric** sorts and punctuation. A text face will sometimes include a number of pi characters drawn in the style of the face; if they are needed for a job, this should be indicated on the **specification** sheet. Additional pi characters are also available in a generic form. Examples of some pi characters are ×, >, and ©. Also known as *odd sort* or **special sort.** *See also* **Pi font.**

Piece fraction. *See* Fraction.

Pied type. Jumbled type.

In handsetting, the type was said to be pied when the letters were returned to the case in a muddle rather than placed in their allotted compartments. Probably derived from "printer's pie," a disorderly heap of type (usually the result of a form dropped by a hapless apprentice). RE

Pi font. A generic font containing multiple **pi characters.** Sometimes both **serif** and **sans-serif** forms are offered, sometimes only serif versions are available. Fonts containing only **ornaments** are also sometimes called pi fonts.

Pilcrow. *See* **Paragraph mark.**

Pin mark. In foundry type, an indentation on one side of the type **body** that carries the name of the founder and the size of the type.

Plain-paper proof. Proof on "plain paper," as opposed to paper with a special surface or photographic paper. *See also* **Dots per inch.**

Today some consider the quality of the plain-paper proofs inadequate for use as camera copy, while others feel they are adequate. The advent of 600 and 1,200 dpi plain-paper machines, coupled with finer toner particles, has greatly improved the quality of plain-paper systems. Nonetheless, a practiced eye can still see a difference between plain-paper output and silver-process output. Moreover, there have been some reports of problems in maintaining consistent density with plain-paper/powdered-toner systems. In all likelihood the advent of electronic prepress will make the matter moot. CME

Plates. Illustrations on pages separate from the text and often grouped into a signature or section called a **gallery.** Derived from the time when illustrations were printed from lithographic plates while the text was printed separately by letterpress. *See also* **Illustration.**

PMT. Abbreviation for *photomechanical transfer* print. A print made by a *diffusion-transfer process,* where the "donor" material is exposed, placed emulsion-to-emulsion with the "receiver" material, and run through a special processor to produce a paper positive that can be used as an **FPO** or as **repro.** *See also* **Photostat; Velox.**

Poetry ellipsis. In verse **extracts,** a line of (usually) em-spaced periods, generally equal in length to the line above and used to indicate the omission of one or more lines or stanzas.

Point. The smallest whole unit in the Anglo-American measurement system used principally in typesetting and printing. The point is equal to 0.35136 millimeter or 0.0138 inch, making 12 points to a pica and approximately 72 points to an inch (actually, 72 points equals 0.9936 inch). *See also* **Didot point; Type size.**

Point size. *See* **Type size.**

Polish hook. *See* **Ogonek.**

Portrait. (1) British term for an **upright** or *normal aspect* illustration, in contrast to **landscape** or **oblong.** (2) A desktop-composition term for a vertical page.

Position print. *See* **FPO.**

PostScript. *See* **Bitmap, bitmapping; Page description language.**

Pound symbol (#). In the avoirdupois system of weights and measures, one pound (equivalent to the abbreviation lb.); also known as a *hatchmark* and used to denote, among other things, number (as a #2 pencil) or space. *See also* **Reference mark; Appendix 8.**

Pound symbol (£). Pound sterling; unit of British currency.

Preface. Part of the **front matter** of a book, usually written by the author to explain the purpose of the work; may include **acknowledgments.** *See also* **Appendix 2.**

Prelims, preliminaries. *See* **Front matter.**

Press-on lettering, press type, transfer lettering. Sheets of individual letters, **ornaments,** or **pi characters,** in a variety of typefaces and sizes, used in preparing **comprehensives** or **camera-ready mechanicals.** The letters are transferred from an acetate carrier sheet by burnishing and are only semipermanent unless sprayed with a protective lacquer.

Press type. *See* **Press-on lettering.**

Primary letter. A lowercase letter that has no **ascender** or **descender,** such as a, e, or o.

Prime character. Full-sized **character.**

Printer's error (PE). Mark placed next to a correction or change in a proof made because of a typesetter's or printer's error, indicating that the correction is to be made at no charge to the publisher. This may be disallowed if the typesetter believes the copy or an instruction was unclear. *See also* **Alteration.**

A PE that is missed by the author and publisher in first proof becomes an AA or EA if it is later corrected. NBP

Printer's fist. *See* **Digit** (2).

Printer's flower. *See* **Flower.**

Printer's query (PQ). A question marked on a proof by the typesetter and addressed to the author, editor, or designer for a decision or response. Queries from the typesetter to the editor or author are usually about the factual accuracy of something in the manuscript (PQ fact), the spelling of a word (PQ sp, or sometimes simply SP?), an apparent inconsistency in spelling, capitalization, or hyphenation, or a point of grammar such as verb-tense agreement. Queries to the designer are invariably about the graphic style of an element. In some cases the typesetter may correct an obvious error, circle it, and mark the notation "OK?" on the proof.

It is the obligation of the typesetter to query only serious matters — indeed, given the time (cost) required, most typesetters would hardly do otherwise. Likewise, it is the obligation of the publisher to acknowledge the queries with either "stet," a change to the proof, or "OK." It has become all too common in recent years for publishers' proofreaders to ignore printers' queries, to write long explanations of why the current treatment in the manuscript is correct, or to tell typesetters in a condescending way to mind their own business. The result of such actions is to discourage typesetters from querying anything, including obvious misspellings. Indeed, some type shops have adopted a policy of no changes or queries, which is hardly to the publisher's advantage. RE, CME, RA, AWS

This works two ways. Publishers can cite instances of typesetters who persistently query or "correct" copy that was perfectly correct in the first place. NBP

Printout. *See* **Hard copy.**

Proof. A copy of set type, disks, or any other stage of manufacture used for checking, correction, or approval. *See also* **Blueline, blue; Galley proof; Page proof.**

Proofmarks. The conventional and accepted series of signs and abbreviations used by **proofreaders** in marking proofs. *See also* **Appendix 8.**

Proofreader. One who reads any stage of **proof** against other **copy** and marks errors for correction.

Pt. Abbreviation for **point.**

Ptr. Abbreviation for printer.

Punch. In typefounding, the original die of a letter or **character,** hand cut or engraved on a hard material, usually steel. The **matrix** used to cast type is made of brass or copper and is formed by striking the steel punch into the metal. A character cast should be an exact re-creation of the image cut on the punch.

Punchcutter. Historically, a person who cut or engraved a **punch** following the designer's drawing for a letter or **character.** The punchcutter sometimes functioned solely as a technician and sometimes collaborated with the type designer on details of the design. Today the term continues to have meaning with *digital punchcutting,* where the type designer's drawings (even those drawn on a computer) must be rendered as computer files. Although more type designers today are probably capable of "cutting their own punches" using computer programs, there is still a need for artisans who can render the image so that it will reflect the design of the type when it is printed.

Quad. Short form for *quadrat,* a nonprinting space **unit.** The term originated in hand composition, where a line of less than full measure must be filled with nonprinting spaces so it can be **locked up.** The sizes available for hand composition are en quad, em quad, 2-em quad, and 3-em quad. For machine composition, *quadding* means adding whatever space is necessary to fill up a line. The terms *quad* and *quadding* are considered archaic but are sometimes still found both as written instructions and as machine commands.

Quad center. *See* **Ragged center; Ragged composition.**

Quadding. *See* **Quad.**

Quad left. *See* Ragged composition; Ragged right.

Quad middle. (1) Composition with **quads** between two groups of words, so that one group is **flush** left and the other flush right. (2) An archaic typesetting instruction placed between elements, indicating that the first is to be flush left and the second flush right. *See also* **Ragged composition.**

Quadrat. *See* Quad.

Quad right. *See* Ragged composition; Ragged left.

Quaint character. *See* Tied letters.

Query. A question marked on manuscript or proof by the editor or the typesetter. *See also* **Carry queries; Printer's query.**

Quotation marks (" "). Marks used to enclose quoted words, phrases, or sentences run into text. United States style uses double quotation marks, with single marks for quotations within quotations. British style often reverses this sequence, and some typographers prefer this style, which allows for more optically even word spacing. In philosophy and linguistics, single quotation marks may be used to enclose a letter or word being discussed. In handset type, raised inverted commas were used for opening quotation marks, and raised commas (apostrophes) for closing ones. This basic character shape is generally followed today – for example, "this" – although a few fonts have opening quotation marks that are mirror images of the closing marks – for example, "this." Unfortunately, typewriter-style quotation marks are also creeping into composition, as in "this." Although most computer composition fonts have double quotation marks as a single character, fine composition frequently requires using two **kerned** single marks to achieve even, close spacing.

Quote marks. *See* Quotation marks.

r. Abbreviation for **recto.**

Rag, ragged. *See* **Ragged composition.**

Ragged center. Composition in which lines are centered, with both left and right margins unjustified; also called *quad center. See also* **Ragged composition.**

Ragged composition. In ragged, or **unjustified,** composition, the **word space** value is kept constant (or almost so), making justification of the lines of type impossible. Thus one edge of the type (or both edges) is allowed to be *ragged* rather than lining up. *See also* **H & J; Justify; Ragged center; Ragged left; Ragged right.**

According to some designers, ragged copy should always be set with a fixed word space without hyphenation. Other designers prefer to allow hyphenation and slight variation in word space. RH

Ragged left, rag left. Composition in which all lines align at the right and **word spaces** are kept constant so that the left margin is uneven; also called *flush right, ragged left; quad right. See also* **Ragged composition.**

Ragged right, rag right. Composition in which all lines align at the left and **word spaces** are kept constant so that the right margin is uneven; also called *flush left, ragged right; quad left. See also* **Ragged composition.**

Raised initial. A **display**-size **capital** letter set at the beginning of the first line of type on an opening page or at a major space break. The base of the initial properly aligns with the **baseline** of the first text line so that the initial is raised above the text line. Also called *stick-up initial* and contrasted with **drop cap.** *See also* **Decorative initial.**

Ranged left. British term for **ragged left.**

Ranged right. British term for **ragged right.**

Ranging figure. *See* **Lining figure.**

Raster image processor (RIP). *See* **Imagesetter.**

Recast. To rerun the hyphenation and justification program on a job that has already been set, usually when going from one proof stage to another. *See also* **H & J**; **Rough pages.**

Recto (r). (1) A right-hand **page** of a book, always odd numbered, as distinguished from a **verso** or left-hand page. (2) The front side of a single **leaf** or sheet.

Reference list. *See* **Bibliography.**

Reference mark. Superior **character** or symbol used within the text, instead of a **superscript** numeral, to refer the reader to a footnote at the bottom of the same page, preceded by the same symbol. The customary sequence is **asterisk, dagger, double dagger, section mark, parallels, number sign** (*, †, ‡, §, ‖, #). *See also* **Footnote call.**

Register marks. Marks used on a **mechanical** for each **overlay** layer as a guide to the printer for its proper positioning. At least three register marks are required on a mechanical. The simplest one is *crosshairs* (+), which should be duplicated accurately on all overlays in the same size and position. These are available on transparent tape strip rolls; more complicated register marks are available as **press-on lettering,** or they may be accurately hand drawn.

Registration. The proper alignment and "fit" of all elements. If the registration is poor, there may be gaps between color areas, overlapping of colors, misalignment of graphics, etc.

Relative unit. *See* **Unit.**

Remake. A **page** that contains enough alterations so that lines must be either added or deleted, causing the page to break differently from the way it did on the preceding proof. Accordingly, the typesetter has to "remake" pages to accommodate the change in the number of lines.

Repro. Short form for *reproduction proof* or *reproduction paper;* the final, **camera-ready** version of any copy. The term originated when composition was with metal type but the printing was to be offset. With this process the final proofs are carefully pulled, sometimes using a special paper, and these reproduction proofs are used as camera copy.

Reproduction paper. *See* **Repro.**

Reproduction proof. *See* **Repro.**

Reproduction-quality print. An image suitable in quality for reproduction. *See also* **Photostat, stat; PMT; Velox.**

Resolution. The number of lines or dots (per unit of measurement) used to form an image. For computer monitors and photocomposition in the United States, this is lines per inch or **dots per inch.** Theoretically, the more lines or dots per inch, the higher the quality of the image. As resolution pertains to monitor screens, when the dots per inch is too high – over 84 or so – text-size type can appear rather thin with currently available **fonts.** When the number of dots per inch is too low, the characters will not be well formed. *See also* **Output resolution.**

Reverse. (1) Instruction to the typesetter or printer to make the color value of an element of **artwork** the opposite of the original, as when black type is to print white out of a background. Type can frequently be reversed when it is typeset, or it can be reversed by the printer. (2) A description of artwork or type that is in negative (that is, reverse) form compared with the original copy: the black values are white, and the white values are black. The archaic form of the term is *reverso*. *See also* **Knock out.**

Revised proof. A second round of **proof** in which corrections marked on the first round have been made. A third round of proof is called *second revised proof.* The term *revised proof* always refers to the same kind of proof; that is, page proof is not called *revised galley proof.* *See also* **First revises.**

RF. *See* **Running foot.**

RH. *See* **Running head.**

RIP. *See* **Imagesetter.**

River. An optical path of white space running (more or less) vertically down a page, sometimes caused by excessive space between words. Rivers may be minimized by hyphenation or by moving words from one line to another to reposition **word spaces.**

rom. Short form for **roman** font, used as an instruction on manuscript or proof. *See also* **Appendix 8.**

Roman. The upright style of type, contrasted with **italic**. *See also* Type style.

Roman alphabet. *See* **Latin alphabet.**

Roman numeral. A number formed from roman letters, I, II, III, IV, etc. (or i, ii, iii, iv, etc.), as distinct from an **arabic numeral.**

Rough breathing. *See* **Breathing.**

Rough pages. Page proof that has not been checked for page breaks (and perhaps has not been proofread) and that is supplied in lieu of **galley proof** as **first proof.** Rough pages are sometimes provided by compositors who use automatic computer **pagination** programs; it makes little sense for them to set galleys, since the entire job will be **recast** to accommodate corrections. Before any job is started, the publisher and type shop should agree on what proof is to be expected at each stage.

Rule. A horizontal or vertical line, either typeset, computer generated, or hand-drawn, that can be included in copy as type or added as art. If typeset, the thickness of the rule is usually measured in **points,** the length in **picas.** *See also* **Appendix 6.**

Runaround, runaround copy. Individual lines of text typeset to fit around the uneven contours of an illustration, figure, or display matter. *See also* **Wrap.**

Run back, run down. *See* **Take back, take down.**

Run in. Instruction on manuscript or proof that heading, caption, or other copy is to be continuous, with no new paragraph. *See also* **Appendix 8.**

Run-in paragraphs, run-on paragraphs. A style of composition where new paragraphs do not start a new line but are indicated by a **paragraph mark** (¶) or other ornamental device.

Run-in sidehead. *See* **Sidehead (2).**

Running foot. Book, part, chapter, or section title positioned below the text area, usually on every text page of a book. As with the **running head,** the **folio** may either be designed as part of the running foot or be unrelated to it. Vertical position of the running foot can either be specified in terms of the distance from the bottom of text (which can be variable), be measured **base to base,** or be in a fixed position and measured from top of the type page.

Running head. Book, part, chapter, or section title or any other reference positioned above the text area, usually on every text page of a book. The **folio** may be related to the running head – on the same line, a line above, a line below, etc. – or they may be designed as two unrelated elements.

Running text. (1) A general description of text, synonymous with **body type.** (2) Text without paragraph breaks (*see also* **Run-in paragraphs**).

Run-on chapters. Chapter openings that do not begin a new page but are spaced down from the last text line of the preceding chapter.

Runover. *See* **Turnover, turn.**

S . . . .

Same size (SS, S/S). A sizing instruction used with **camera-ready copy** to indicate that the final size will be the same as the original copy. Marking the copy "100%" is more exact.

Samples, sample pages. Proof generated by the typesetter, usually including a sample of the **repro,** showing representative typographic elements that appear in a book. Samples are usually provided in page form, but sometimes samples in galley form are all that is needed. The role of samples is to assure designers that their ideas, as reflected on the **layouts** and **specification** forms, are correct and that the type shop has understood the specifications and layouts. Following are most of the things that should be shown in sample pages: (1) repeating **display type** – such as chapter openings – to check for size, position, and spacing (including any **letterspaced** material); (2) any special **pi characters** if there is any question as to their availability or appropriateness (such as a **hamza** or a slashed ell); (3) examples of the text elements, including **text, extracts**, lists, **epigraphs, subheads** (including **stacked heads,** if any), **running heads** or **running feet, folios,** and perhaps a **table** if there is sufficient reason; and (4) material that will have to be "constructed" for the job, such as **italic small caps,** which usually have to be **faked.**

Throughout most of the twentieth century sample pages showed only a chapter opening page and a text spread containing frequently occurring elements – things where any change in style after the job had been composed would mean substantial changes to the text throughout the entire job, a potentially disastrous situation. Today many publishers are requesting more extensive samples, including the title page, table of contents, several chapter openings, several back matter headings, etc. Although designers or publishers may feel that these extensive samples are necessary, their preparation is time consuming – hence costly – so there is an increasing tendency for type shops to charge more for them.
CME

Sans. Short form for **sans serif.**

Sans serif. A classification of typefaces that do not have **serifs,** such as Futura. *See also* **Appendix 3.**

sc. *See* **Even small caps; Small caps.**

Scanner. A machine that reads images and converts them to digitized data that can be read by a computer. In book production, a scanner may be used to "read" a manuscript in an attempt to save some of the costs of **keyboarding.** Scanners can also "read" pictures, which can in turn be incorporated into a PostScript file containing both text and art. *See also* **OCR.**

Scansion marks. Marks above or below a syllable that show the rhythmic components of verse. In classical languages these are the **longum** (¯) and **breve** (˘) for long and short syllables, respectively. In English it is the accented syllable, marked by the *ictus* (´), that establishes the pattern. *See also* **Appendix 7.**

Schwa (ə). Phonetic symbol for a neutral vowel, typically occurring in unstressed syllables, for example, the final syllable in the English

word *sofa.* From the Hebrew *sh'wa,* a short vowel usually transliterated as a superior lowercase e.

Scotch rule. A rule constructed of two parallel lines, usually of different thicknesses. *See also* **Double rule.**

Screen. *See* Tint.

Screen font. *See* Bitmap, bitmapping.

Script. Type resembling handwriting, with letters that appear to be
connected. *See also* **Cursive; Appendix 3.**

Scuffed. A term used to describe **repro** that has been damaged.

Search and replace. *See* Global search and replace.

Second half title. *See* Half title.

Second-order character. *See* Subscript; Superscript.

Section. A part or division of a work. *See also* **Part title; Subhead.**

Section mark (§). Mark used to indicate a **section** in a book or as a
reference to a footnote. Frequently used with legal references with
numbered sections. *See also* **Paragraph mark; Reference mark.**

Semibold. A stroke-thickness variant of a typeface, heavier than normal or "text weight" **roman** and lighter than **bold.**

Series title. Usually the first page of a book that is part of a series.
Alternatively the series title may appear on p. ii of the **front matter,**
often with the name of the series editor and a list of the titles
already published in the series. *See also* **Appendix 2.**

Serif. A terminal or short ending stroke of a type character. *See also*
Sans serif; Appendix 1.

Set, setting. (1) Short term for *typeset* or *typesetting,* as in "set 10/13 ×
26." (2) A reference to the character of type, as in "Perpetua is too
thin in a 10-point setting" or "Caslon (metal) has a rather open
setting." *See also* **Set width.**

Set solid. Describing lines of type that have no extra space (**leading**)
between them, for example, 10/10.

Set width. In computer composition, the set width is the amount the
typesetting machine escapes when instructed to set a letter (*see also*
Escapement). More generally, set width is the horizontal space a
character takes up regardless of what is shown when it is actually

printed; that is, the space taken up by the character plus the left and right **sidebearings.**

SGML. *See* **Standard generalized markup language.**

Shaded letter. *See* **Shadow letter.**

Shadow letter. A group of typefaces having a shadow effect on one side or more of each letter.

Sheet. *See* **Leaf.**

Shilling fraction. *See* Online **fraction.**

Shilling mark, shilling stroke. *See* **Slash.**

Short descender. The portion of a **lowercase character** that falls below the **baseline** and is designed to be especially short. When short **descenders** (and **ascenders**) are combined with a large **x-height,** such typefaces can be set rather small, with minimal **leading,** and still be quite legible. This is a desirable feature when setting some kinds of printed material (newspapers, for example), since it allows copy to be set closer. With both foundry type and Linotype, some **fonts** are available with both **long descenders** and short descenders, and their use depends on the job at hand.

Short page. A page that has been made up shorter than the specified page depth in order to avoid a **bad break.** *See also* **Long page.**

Short-ranging figures. *See* **Small cap figures.**

Shoulder. In metal type, the lower, nonprinting surface on the top of the type body at the base of the type character or image. It provides space for the descending strokes of letters such as g, j, q, and y.

Show-through. *See* **Backing up** (2).

Sidebar. A column on the **type page** that is outside the **text page,** sometimes used for reference notes or for **captions.** Sidebars are sometimes designed to extend into the main text area. *See also* **Sidehead; Side note.**

Sidebearing. The spaces on each side of a letter that are part of the physical space it occupies (its **set width**). The sidebearings determine both the **character fit** within a font and whether it has a close or loose setting. If the sidebearings are not carefully and evenly rendered by the type founder, the character fit will be poor, with some letter pairs being too tight and others too loose. To some extent this condition can be overcome by **kerning.**

The total amount of sidebearing space relative to the space occupied by the characters determines how closely the letters fit together. For example, the sidebearings of Galliard – indeed, most ITC-developed fonts – are rather small, whereas Linotron Granjon has rather large sidebearings. Sometimes mechanical factors can affect sidebearing relationships. For example, fonts in text sizes developed for the Linotype linecaster frequently had the same set width for the italic letters as for the roman (since they could be on the same physical matrix). The result of these large sidebearings was that unless the italic was recut, it looked rather gappy. (The recuttings had their own problems; the traditional slenderness of italic fonts was lost.) Most photocomposition systems allow the overall setting (close vs. loose) to be modified by tracking or white space expansion or reduction, although such features generally affect only the left sidebearing. With the newer PostScript fonts, programs are available that let the user modify the type itself, including the sidebearings. CME

Side folio. A page number that is positioned to the side of the **text** block, usually in the **outside margin.**

Sidehead. (1) A **subhead** placed at the side of the main **text,** in the **outside margin.** (2) A heading set at the beginning of a paragraph, usually in a different font or style, and followed by a period; also called *run-in sidehead.* (3) In **tables,** a subhead that is wholly or partially in the **stub,** that is, placed with the left-hand column entries in the table body. Table sideheads may have extra space above and below to set them off from other stub entries, or stub entries may be indented from the sidehead. Finally, even though a sidehead begins in the stub, it may be allowed to run into the body of the table. *See also* **Subhead; Appendix 6.**

Side note. A supplementary note that usually appears in the **outer margin** of a page or in a **sidebar** column.

Silverprint. *See* **Blueline, blue.**

Single quotes. *See* **Quotation marks.**

Sink, sinkage. The area from the top of the **type page** (not from the trim) to the first type element on that page. In **photocomposition**, preferably given in **points** and **picas** to the **baseline** of the first type element; in metal composition, given to the top (**cap** or **ascender**) of the first type element.

As with so many specifications, the convention of giving sinkage from the top of the type page came about because of the way books were manufactured with metal type and letterpress printing. In the future the convention may change; for example, it may turn out to be easier for all concerned if sinkages are given from the trim. The point is that conventions of this sort are adopted to ease manufacturing, not because they are "right" in any absolute sense. CME

Sink initial. *See* **Drop cap.**

Slab serif. *See* **Egyptian; Appendix 1; Appendix 3.**

Slant. *See* **Slash.**

Slanted roman, sloped roman. *See* **Oblique.**

Slash (/). A diagonal line or oblique stroke sloping downward from right to left. There are two **sorts** that have this characteristic. The first of these has as its **set width** the full width of the character, as |/|, where the vertical rules show the set width. The second, И, has a set width less than the full width of the character. The first is always termed a *slash* or *virgule,* the second a *fraction bar.* Various authorities use the terms *solidus* and *shilling stroke* or *shilling mark* with either. For example, *The Chicago Manual of Style* equates solidus and slash; Robert Bringhurst (*see* Bibliography) equates solidus and fraction bar. Linotype, in its *Linotron 202 Font Handling User's Guide,* equates the slash with the shilling stroke; Bringhurst equates the shilling stroke with the fraction bar. Only the terms *virgule* and *fraction bar* are never equated; for this reason they are perhaps the clearest. With Postscript Adobe Standard Encoding, the terms (names) used are *slash* and *fraction.* One common use of the slash is to indicate line breaks when poetry is run into the text. Two thin-spaced slashes indicate omitted lines (//), and one slash indicates a line break (/). For example: "I'm truly sorry man's dominion / Has broken Nature's social union // The best-laid schemes o' mice an' men / Gang aft a-gley." A **backslash** (\) slopes downward from left to right. *See also* **Fraction.**

Many publishers specify the space on each side of the slash, but if that space is put in mechanically rather than optically, the result is usually marked "PE." CME

Slashed ell (Ł, ł). Character used in Polish, Navajo, and other languages. Also known as *barred ell.*

Slashed oh (Ø, ø). Character used in Danish and Norwegian. Also known as *barred oh.*

Sloped roman. *See* Oblique.

Small cap figures. Also called *Short-ranging figures.* Numerals of uniform height, extending very little beyond the **x-height.** Small cap figures are usually offered as a supplemental character set for use with **small caps** when **old-style figures** are not available. Examples include Linotron Aster, Century Expanded, and Century Schoolbook.

Small caps, small capitals (sc). **Capital** letters drawn to approximately the same height as the **x-height** of a **font.** Since the proportions of small capitals are slightly different from those of capital letters, and since small capitals are designed to have the same stroke weight as other characters in a font, setting the normal caps of a typeface in a smaller size does not give the same aesthetic results as using designed or **true-cut** small caps. As with all capital letters, a small amount of **letterspacing** generally improves the appearance of small caps. Should numbers be used within a small cap setting, **old-style figures** (or small-cap-height **lining figures**) should always be used, for example, ELECTION OF 1992, never ELECTION OF 1992. *See also* **Fake small caps; Machine italic.**

Smooth breathing. *See* Breathing.

Soft hyphen. A special type of hyphen in a computer file that indicates where a word may be hyphenated in order to break a line. The soft hyphen (also called *discretionary hyphen*) is a different character from the regular hyphen (**hard hyphen**) and is "soft" in that it will disappear if – as a result of editing or reformatting the file – the hyphenated word no longer falls at the end of a line. When manuscripts prepared on word processors are converted to files for typesetting, soft hyphens can be globally deleted while the hard hyphens are left alone.

Soft return. In a computer file, an end-of-line command that converts to a normal word space when a document is reformatted and is different from the end-of-paragraph return (**hard return**). As with the **soft hyphen,** preserving the two types of returns is important when converting an author's word-processing files for typesetting.

Solid. *See* **Set solid.**

Solidus. A **slash.** This term is frequently used when referring to an oblique stroke in a mathematical context.

Solidus fraction. *See* Online **fraction.**

Sort. A piece of type, or **character,** of a specific **typeface, font,** and size. *See also* **Special sort.**

SP. Proofreaders' mark, generally circled, meaning *spell out. See also* **Appendix 8.**

SP? Spelling **query** on manuscript or proof, addressed by the editor to the author or by the typesetter to the editor. *See also* **Appendix 8.**

Spacebands. Expandable metal wedges added between words in a line of matrices on a **linecaster.** Several ranges of spacebands are available, with the **thin space** being preferred for bookwork. The term is still used by some designers and typesetters with photocomposition equipment for **word spaces** or word-spacing parameters. *See also* **Fixed space.**

Space symbol. *See* **Pound symbol** (#).

Spanner head. A heading in a **table** that "spans" or "straddles" at least two other heads, either **column heads** or other spanner heads; also called *crosshead* or *straddle head. See also* **Appendix 6.**

Spanner rule. In a **table,** a horizontal **rule** under the **spanner head** that shows the range it covers; also called *cross rule* or *straddle rule. See also* **Appendix 6.**

Special character. *See* **Special sort.**

Special sort. (1) An uncommon **character** specified to be in the same typeface as the rest of the text. For example, a dot over a lowercase t can be made up from other characters in the font – the period (or centered dot) and the lowercase t. If special sorts are needed for a job, they should be listed on the **specification** sheet. (2) A synonym for *odd sort* or **pi character** when the required character is not available in the specified type style. *See also* **Appendix 7.**

Specifications. Complete instructions for typesetting and positioning of every element of a job. Usually referred to as *specs. See also* **Appendix 4.**

Specs. Short form for **specifications.**

Spiritus asper. *See* **Breathing.**

Spiritus lenis. *See* **Breathing.**

Spread. Facing pages of a book, regarded as a single entity. Also called *double-page spread* or *two-page spread.*

Square brackets. *See* **Brackets.**

Squared-up. Referring to an image shaped, by cropping if necessary, to a perfect rectangle.

Square up. Instruction to adjust the position of an element on a page so that it is ninety degrees from the vertical or horizontal.

SS, S/S. *See* **Same size.**

Stacked heads. One level of **subhead** following another with no text in between.

Standard generalized markup language (SGML). A highly structured language used for **formatting** documents that is not specific to any particular computer or software and thus allows a document to be transferred between computers or programs while keeping the formatting intact.

Standing type. Term originating with metal composition, referring to composed type stored at the type shop after platemaking or printing. If there is no standing type, type has to be entirely recomposed. In **photocomposition:** (1) type set in galleys for a journal that is to be held for use in a future issue – either a cover or an article that is not to be included in the current issue; (2) a copy of the electronic "files" used for setting a job that is to be saved for future use. *See also* **Keep standing.**

Stat. *See* **Photostat, stat.**

Stem. The principal or thick stroke of a letter. Also called *main stroke. See also* **Appendix 1.**

Stet. An instruction to ignore a change requested or marked, to let copy stand as it appears in manuscript or proof; from the Latin for *let it stand.* Material to be stetted is indicated by a row of dots underneath the character, word, or phrase. *See also* **Appendix 8.**

Stick-up initial. *See* **Raised initial.**

Straddle head. *See* **Spanner head.**

Straddle rule. *See* **Spanner rule.**

Straight copy. *See* **Clean** (1).

Stress. One of the classificatory properties of a **typeface;** the orientation of a line drawn through the thinnest parts of the letter. With Bodoni, for example, the stress line is vertical. With Bembo it slopes downward left to right at about thirty degrees. *See also* **Appendix 1.**

Strike-on. Describing a form of composition in which the **repro** is generated by a "typewriter" using a carbon ribbon on special paper, usually a clay-coated stock. IBM, A-M, and Compugraphic all developed machines for strike-on composition.

Strip-in correction. A correction to **camera-ready copy** that is pasted or stripped in by hand on the original page, as opposed to resetting an entire page with the corrections incorporated. The advantage of strip-in corrections is that only the reset patch needs to be proofread. Care should be taken to match the **density** of the original type.

Stripping guides. *See* **Corner marks.**

Stroke. *See* **Main stroke; Stem; Appendix 1.**

Stub. The left-hand column of a **table** when it includes guiding entries. *See also* **Appendix 6.**

Style as. Term used in type specifications, usually followed by *but.*

For example, a bulleted list may be specified as "style as numbered list, but use two-point bullet in place of the number."

Style sheet. (1) A set of guidelines supplied by a typesetter or a publisher that defines **generic codes** used or defines how individual elements in typesetting and paging are routinely handled; also called **house style** (*see also* **Appendix 5**). (2) A sheet supplied by an editor showing the specifics of editorial style for a particular book that either are not covered by general convention or vary from convention.

Subhead. Section title or heading subordinate to the chapter title. The primary order of subhead in a text is marked on the manuscript or proof with the code letter A (**A-head**). The second order of subhead is coded B, the third C, and so on. Subheads may also be coded numerically, with 1 equivalent to the primary or A-head. Typographically, their design should reflect their relative importance. *See also* **Appendix 4.**

Subscript. A letter or number smaller than the main text size and set below the **baseline** of standard text – either about 2 points lower with a 10-point setting or with the subscript baseline aligned at the bottom with the **descender** of a text character. The term is most frequently encountered in mathematical setting. Also called *second-order* or *inferior character.*

Subsubscript. A **subscript** to a subscript. *Supersubscript* and *subsuperscript* characters are also possible. Normally encountered only in mathematical setting, these are more commonly referred to as *third-order characters* (super- and subscripts being *second order* and full-size characters being *first order* or *prime*).

Sunken initial. *See* **Drop cap.**

Superior. *See* **Superscript.**

Superscript. A letter or number smaller than text size and usually aligned at the top with the text type **ascender,** used to refer to a **footnote** or **endnote** or in mathematical setting. Also called *second-order* or *superior character.*

Supersuperscript. A **superscript** to a superscript. *See also* **Subsubscript.**

Swash letter. A letter (usually found in **uppercase italic**) with a decorative **flourish,** for example, *A* and *N.* **Lowercase** swash letters, such as *w,* occur infrequently. *See also* **Appendix 1.**

Swash terminal. *See* Terminal (3); **Appendix 1.**

Swelled rule. *See* Tapered rule.

System. Short form for *front-end system;* can also be used to mean the computer program in a type shop.

T . . . .

Tab. Short form for *tabulate, tabulation.* (1) To set type in a series of columns (*see also* **Table**). (2) A keystroke that creates one or more preset widths of indention in a typewriter or word processor file.

Table. Information organized in a series of columns and rows that relate horizontally, vertically, or both. Although many of the complexities of table design can be resolved only by a skilled compositor, the publisher should supply a manuscript that is organized, concise, consistent in style, and cleanly presented. To achieve this may require retyping. Tedious though this may be, the eventual saving in time and typesetting costs can be considerable. *See also* **Appendix 6.**

Tabulation. *See* Tab.

Tagline. *See* Galley.

Tail. (1) The bottom trimmed edge of the page. (2) Often used to refer to the tail or **bottom margin.** *See also* **Margin.**

Tail margin. *See* Bottom margin; **Margin.**

Tailpiece. An **ornament** or small illustration used as a **flourish** at the close of a chapter.

Take back, take down. A marginal instruction, used in proofreading, to transpose a specified portion of text back to the preceding line or forward to the next line; also called *run back* or *run down. See also* **Transpose; Appendix 8.**

Tapered rule. A **rule,** usually shorter than the type measure, that is tapered at each end; also known as a *swelled rule.* Tapered rules are commonly used in centered typography as a decorative device or separating element.

Tearsheets. Pages removed from a book or an unbound signature, used by suppliers of typesetting, printing, or design as samples for prospective clients or for reproduction in exhibition catalogs, etc. Tearsheets are often supplied in place of typescript for typesetting a new edition or parts of a work that has been previously published. Since these are often less legible than manuscript, particularly when editorial changes have been made between the lines of type, the typesetter may impose a penalty fee. *See also* **Penalty copy.**

Terminal. (1) Short for *video display terminal* (*VDT*), a cathode-ray tube that displays text as it is keyboarded or read from a tape or disk. The operator is able to edit the text on screen, check and change style codes, and (depending on the system) display complete pages of text including footnotes and folios. One should remember, however, when making requests of the compositor, that the screen image is not an exact rendering of the final type output. Subtle refinements in visual letterspacing, for example, may require considerable trial and error. (2) The end stroke of some letters. (3) An alternative **lowercase** letter designed for use at the end of a word to impart a concluding **flourish;** also called a *swash terminal* (for example, &). *See also* **Appendix 1.**

Text. (1) The main matter of the page, as distinct from specific elements such as **chapter heads, subheads,** or **extracts.** (2) The body of a book, as distinct from **front matter** and **back matter.** (3) Written matter as distinct from **illustrations** or **tables.**

Text area. *See* Text page.

Text font. The **typeface** used in setting the **text,** as distinguished from the **display.** The same face may be used for both, hence the instruction to "set display in text font."

Text page. The area of the **page** defined by the main **text,** excluding **running heads, drop folios,** or marginalia. Thus, in the type specifications the text page might be defined as "38 lines of Baskerville 10 on 13 pt justified to 24 picas." It may also be expressed in **picas** and **points,** measured from the **ascender** height of the first line of text to the **baseline** of the last line of text. *See also* **Type page.**

Text type. *See* **Body, body type.**

Thick. *See* **Thick space.**

Thick space. Often referred to as a *thick* and equal to one-third of an em. Thus, in a line of 12-point type a thick space would equal 4 points, or 6 units in an 18-unit system. It may also be expressed as a ⅓ em, 3-to-em, or M/3. *See also* **Unit.**

Thin. *See* **Thin space.**

Thin space. Often referred to as a *thin* and equal to one-fifth of an em or 3.6 units of an 18-unit em. Thus, in a line of 12-point type a thin space would equal approximately 2½ points. It may also be expressed as a ⅕ em, 5-to-em, or M/5. It should be noted that several authorities, among them *The Chicago Manual of Style,* disagree with this definition, giving the thin a value of 4.5 units (4-to-em), and that some compositors using computer equipment set their thin spaces smaller than 5-to-em. *See also* **Hair space; Unit.**

Third-order character. *See* **Subsubscript.**

Thorn (Þ, þ). Character used in **Old English,** Middle English, and Icelandic to represent the unvoiced th.

Three-em dash. *See* **Dash; Em dash.**

Thumb margin. *See* **Outside margin.**

Thumbnail. A miniature sketch used especially in the design of illustrated books to plan the **layout** of a sequence of pages.

Tie. *See* **Tied letters.**

Tied letters. Generally taken as synonymous with **ligature;** two or more letters that are combined as a single **sort,** for example, Æ, æ, and Œ, œ. Sometimes used more specifically to describe ligatures joined by a visible tie stroke, such as ȸt; also called *quaint character. See also* **Appendix 1.**

Tilde (˜). Diacritic used above n in Spanish and above vowels in Portuguese. *See also* **Appendix 7.**

Tint. Also called *screen.* A tone, usually expressed as a percentage of a solid color. For example, to render the display type on a title page as a middle gray, the typesetter (or printer) would be instructed to "reduce type to 50 percent tint (or screen)."

Tissues. The designer's page **layouts** that form the visual part of the specification to the typesetter, so called because traditionally they were drawn onto semitransparent vellum paper suitable for tracing letters from alphabet sheets. Layouts produced by computer and supplied as laser prints or photocopies on white bond paper are now sometimes called **mactissues.**

Title page. Part of the **front matter** of a book that gives the full title, the name(s) of the author(s) or editor(s), and the publisher's **imprint.** *See also* **Appendix 2.**

Titling font. A **typeface** of **capital** letters only, occupying virtually the full depth of the **body** and thus larger than a text face of the same nominal size. More generally, one of a **family** of fonts that has been drawn specifically for use in larger sizes but remains in harmony with the related **text font.** Stanley Morison's Times Titling is a distinguished example from the past. More recently, Robert Slimbach's Minion series includes a titling font subtly modulated to look good in display sizes of 18 points and up while retaining the **color** of the text. *See also* **Display face.**

Top margin. *See* **Head margin.**

Top of text page. The top of the **ascenders** of the first line of the main text, without reference to the **running head.** This term is sometimes used to define the point from which illustrations should hang on the page, so that they retain a visual relation to the text block. *See also* **Text page; Top of type page.**

Top of type page. The **ascender** height of the topmost element on the page, often the **running head.** *See also* **Type page; Top of text page.**

Track, tracking. The subtraction of space between letters applied to the entire alphabet. Some type shops, particularly those that serve advertising agencies, offer a choice of **letterspacing** styles, or tracks, which are preset modules in the computer program. The higher the track number, the tighter the letterspacing. The actual space value (in points or ems) signified by the track number depends on the typesetting system employed. Commonly, track 1 = medium, track 2 = tight, track 3 = very tight. Tracking is intended primarily for large **display** type. *See also* **Kern; Minus letterspacing.**

Generally it is courting disaster to interfere with the type designer's conception in this way. Kerning, the selective tightening of particular combinations of letters, is entirely another matter. RE

For text sizes of most typefaces, even track 1 appears too tight. More subtle tightening is employed as a matter of course in good bookwork to compensate for the tendency of a typeface to appear loosely set in larger sizes, usually 15 point and up. There are, however, a few typefaces that may benefit from a slight tightening or slackening of overall letterspacing in text sizes. CME

Transfer lettering. *See* **Press-on lettering.**

Transitional typefaces. *See* **Family.**

Transpose (tr). Instruction on manuscript or proof to reverse the order of two letters, words, sentences, etc.; to move copy from one place to another. Sometimes expressed as *tr up, tr down,* or **take back, take down.** *See also* **Appendix 8.**

Trim, trim size. (1) The finished size of a **page** after the edges have been trimmed. (2) The trimmed edges of the page, as in *head trim, foot trim,* or *fore-edge trim. See also* **Bleed.**

Trim marks. Guidelines printed at the corners of the page to indicate where it should be trimmed. *See also* **Corner marks.**

True-cut. A variation of a **typeface** that has been specifically designed to be part of a type **family** and is not machine generated. *See also* **Cut; Fake small caps.**

TrueType. A computer file format, similar in many ways to PostScript, for setting type.

Truncated setting. Justified setting in which words are broken willy-nilly and hyphens are omitted to achieve even interword spacing; a device used in type sample books and occasionally in concrete poetry.

Turn. An instruction to rotate an element so it appears sideways on the page. *See also* **Broadside.**

Turnover, turn. Also called *runover*. Lines of an entry that continue over from the first line. For example, in specifying the style for an index, one might ask that turnovers be indented one em.

Two-page spread. *See* **Spread.**

Two-up proof. A proof, often in the form of a photocopy, that shows two **facing pages** on each proof page. The margins may be cut short to accommodate the paper size of the copying machine. If the customer needs to see the book page with margins intact, a **one-up proof** may be required.

Type area. A very ambiguous term when applied to bookwork. Some use it to mean **type page,** some to mean **text page.** It is therefore best avoided.

Typeface. One of the variations or styles in a type **family.** The design of a type family, including its shape, weight, and proportions, makes it distinct, but it usually exists in many sizes. *See also* **Appendix 3.**

Type family. *See* **Family.**

Type founder. Originally, the owner of a type foundry, a manufacturer of metal type for **handsetting.** This sense is retained in the term *founder's type* or **foundry type.** Now used more generally for any maker of type, whatever the process of manufacture.

Type gauge. Also called a *type scale, pica pole, pica rule, pica stick.* The simplest form is a measuring device like a ruler, usually calibrated on one edge in **points** and **picas** and on the other in inches. Some are printed on a transparent plastic sheet. Some type gauges are slotted and may include many calibrations in various units of points, from 5 points up to 14 points. The best known of the slotted gauges is the Haber rule, whose name has become almost generic. As far as can be ascertained, there is no type gauge generally available in the United States that offers calibrations in half-point increments, a serious deficiency.

Type markup. The marking of a manuscript for typographic style by a designer or production editor. *See also* **Appendix 4.**

Type measure. The width of a full line of type, usually expressed in **points** and **picas**.

Type page. The entire area to be printed, including the **running head** or **running foot** and the **folio**. It is measured vertically from the **ascender** of the first element to the **baseline** of the last. On pages with **drop folios**, specifications should make clear how this element figures in the measurement. The width of the type page also includes any marginalia such as **sideheads, sidebars, side notes,** or **side folios.** The page size is specified in **picas** and **points,** with the width always preceding the depth (the opposite of British practice). *See also* **Text page.**

Type scale. *See* **Type gauge.**

Typescript. *See* **Manuscript.**

Typesetter. (1) A person who sets type, either the keyboard operator (who is essentially a typist) or the **compositor** (who is responsible for interpreting the type specifications and encoding the manuscript). Matters of style are usually addressed specifically to the compositor, or *comp.* (2) A typesetting machine.

Typesetting. *See* **Composition.**

Type size. The size of a specific **font.** Until late in the nineteenth century, type size was indicated by name (**agate,** for example, corresponding to approximately 5½ points). This was an imprecise convention, since the size denoted by a particular name varied from one type foundry to another. In 1878 the **point** system of measurement was introduced in the United States, and it has since become standard throughout the English-speaking world for a variety of measurements including the size of the type itself. The point size of a typeface is not a simple measurement of the letterform alone but pertains to the **body** that carries it. This is most easily understood with reference to metal type, for which the point system was first devised. The body – the block of metal that carries the letter – includes a space below the letter itself, called the **beard.** Because the point size includes the beard space, it is somewhat larger than the actual letterform. Depth of letterform varies from one typeface to another: for example, this is 11-point Centaur, and this is 11-point Bookman.

Type style. The formal characteristics of a **typeface.** (1) Used in a broad sense to identify roman, **bold roman,** *italic, **bold italic,*** condensed, **expanded,** etc. (2) Used in the sense of type **family** or group. *See also* **Appendix 3.**

Typo. Short form for a typographical error made by the typesetter, a **printer's error.** *See also* **Alteration.**

Typographer. Formerly a typographer both designed and set type. Today the term is loosely used to refer to the **typesetter,** to the typographic designer, and occasionally to someone who designs **typefaces.**

Typographical error. *See* **Printer's error.**

Typography. The arranging of type in an appropriate manner to suit a particular purpose.

Typositor. Short for the trade name Photo Typositor, a photographic typesetting system that projects type patterns, letter by letter, through a negative filmstrip wound between two spools. Interchangeable lenses permit various forms of distortion: expanding, contracting, or angling of the letterforms. Very much a hand operation, this is a slow process but is capable of sharp **character** definition and refined **letterspacing.** For these reasons it is used primarily for **display** setting. High-resolution phototypesetters have to some extent superseded the Photo Typositor, but it is still used.

U

UC. Abbreviation for **uppercase** or **capitals.**

Ulc, U/lc, U&lc. Abbreviations for *upper- and lowercase. See* **Caps and lowercase.**

Umlaut (¨). Diacritic placed over a, o, or u in German, indicating omission of an e following and affecting the pronunciation of the vowel. *See also* **Diaeresis; Hungarian umlaut; Appendix 7.**

Uncial. A rounded form of **capital** letter or **majuscule** first used in Latin and Greek texts about the third century and later adopted by medieval scribes.

Underdot (̣). Diacritic used in romanized Arabic, Hebrew, Sanskrit, and some African and Native American languages. *See also* **Appendix 7.**

Underline. *See* **Underscore.**

Underscore. (1) In typesetting, a fine **rule**, set just below the **baseline** and cutting through the **descenders,** used as a mark of emphasis. (2) In marking manuscript or proof, a single line drawn below a word to indicate that it is to be set in **italic.** *See also* **Appendix 8.**

Undotted i. *See* **Dotless i.**

Undotted j. *See* **Dotless j.**

Unit. A measurement, always a fraction of an **em** (and therefore relative to **point** size), used to define the width of typographical elements and spaces. The width of a letter so defined refers to its **body** width rather than its visual width and thus includes the fixed spaces on each side of the letter that are part of its design and determine its relation to other letters in a word (the **sidebearings**). The unit value varies from one typesetting system to another and is a constant source of confusion. The Monotype Corporation at present uses 96 units to the em for filmsetting, whereas its metal typecasters employ an 18-unit system. The Linotype Corporation's Linotron 202 phototypesetter uses two systems: 54 units to the em for **letterspacing,** but 18 units to the em for **word spacing.** Programs using PostScript have adopted the decimal system, with 1,000 units to the em being most common. *See also* **Body, body size.**

The person responsible for specifying type style is confronted by the unit system when attempting to define spaces between words or letters. Some designers attempt to avoid confusion by describing letterspacing in terms of points and word spacing as tight, regular, or loose. This is hardly satisfactory, however, since the point, except when used with reference to large display type, is too large a unit to allow for subtle adjustments, and the terms tight, regular, and loose, though expressing an agreeable optimism, are entirely too vague. One way to minimize confusion is to define in the type specifications the system employed; for example, "All word- and letterspacing expressed in 54 units to the em." The best practice is to ask for the most appropriate method of specification for the particular typesetter's system. Institutions like state university presses, which are usually required to obtain competitive bids from several potential suppliers, should be particularly wary in this regard, since each bidder may employ a different system and so interpret the specification differently. Some European type manufacturers becoming established in the United States, most notably the Berthold Company, base all measurements on the millimeter. This system is well explained in the company's type catalogs, and it has a logic that recommends its universal adoption. Typesetting on personal computers is in a state of evolution, and systems of measurement vary with each software program and from one generation to another of the same program. RE

Universal search and replace. *See* **Global search and replace.**

Unjustified. Descriptive of typesetting in which line lengths are not aligned but ragged. In unjustified setting the word spacing is constant (or varies only slightly) and the line length is variable. *See also* **Ragged composition.**

Updated disk. *See* **Editorially correct disk.**

Upper- and lowercase (Ulc, U/lc, U&lc). *See* **Caps and lowercase.**

Uppercase (UC). Capital letters. *See also* **Case.**

Upright. (1) Descriptive of a page format, table, figure, or illustration that is taller than it is wide; also called **portrait** or *normal aspect* as opposed to **broadside** or **landscape.** (2) Descriptive of letters with vertical **stress,** as in most **roman** fonts (*see also* **Appendix 1**).

V . . . .

v. Abbreviation for **verso.**

Vandyke. *See* **Blueline, blue.**

Velox. A high-contrast photographic contact print of a **halftone** or screened image, typically of excellent sharpness and density. Used as a position print (**FPO**) or for **camera-ready artwork.** Veloxes are generally made by the printer's cameraman. *See also* **Photostat; PMT.**

Venetian. A category of **typefaces** sharing particular attributes of **stress** and stroke weight, often included in the larger **family** of *old face. See also* **Appendix 3.**

Verso (v). (1) A left-hand **page** of a book, always even numbered, as distinguished from a **recto** or right-hand page. (2) The back side of a single **leaf** or sheet.

Vertical rule. A typeset line that runs from top to bottom on a page, not from left to right (a *horizontal rule*). Used primarily in setting tabular matter, though it is often an impediment rather than an aid to clarity. *See also* **Rule; Appendix 6.**

Virgule. *See* **Slash.** Not to be confused with French *virgule* (comma).

Visual space. (1) The space between elements (words, letters, lines of type, rules, etc.) apparent to the eye (optical space), as opposed to **machine space.** For example, in the letter combinations LY and IN, the disparity of visual space within the two pairs is obvious, but mechanically the spaces are equal. (2) Within the typesetting business, the term also refers to the measurable gap between the **descenders** of one line and the **ascenders** of the next. If the first line lacks descenders, the measurement is from the **baseline.** *See also* **Optical center; Optical letterspacing.**

Visual spacing. *See* **Visual space.**

Vortex. The bottom junction of two **stems,** as in the letter V. *See also* **Appendix 1.**

Vowel ligature. Vowels combined into a **diphthong** and set as a single sort (Æ, æ, Œ, œ).

Wedge. *See* Háček; Spacebands.

Wen. *See* Wyn, wynn.

WF. Abbreviation for **wrong font.** *See also* **Appendix 8.**

White space. (1) In **photocomposition,** usually taken to refer to the space between letters and words, which can be affected by *white space reduction* routines. (2) More generally, the unprinted areas of a page, such as the **margins,** *interline space,* **sinkage,** and indention. *See also* **Minus letterspacing.**

White space reduction (WSR). *See* **Minus letterspacing.**

Widow. A single word or short line at the head of a page. The term tends to be used loosely and is often confused with **orphan** (a single word on a new line or a very short line at the end of a paragraph, especially at the foot of a page).

A widow is generally deemed persona non grata and is shunned at whatever inconvenience. A short widow, although the most offensive, is also the easiest to deal with. It is usually dispatched by an adjustment to word spacing or page depth. If this is not possible, the editor may be asked to eliminate it by rewording the text. A long widow is sometimes spaced out, but the result is likely to be gappy. Remember that in getting rid of a widow you may be landed with an orphan, or even several orphans – and quite possibly with another widow as well. Therefore their eradication is best left to a professional, the compositor generally having the most experience in these matters. One sort of widow often tolerated, and even taken for granted, is the unattached widow – that is, a subhead or a line of dialogue, such as "Ah!" These lines are not necessarily thought of as widows, but they behave like widows in disrupting the unity of the page, and they also may make running heads look awkward. RE

Widows are most often removed by the compositor either by running the affected pages long or short or by making or losing a line (in effect, a change of word spacing) in some other paragraph on the affected spreads. CME

Window. A solid black area inserted by the typesetter (or Rubylith film or other opaque material added to **repro**), representing the area occupied by an **illustration.** Generally speaking, windows have been supplanted by **FPO**s.

Wood letter. Wooden type manufactured for letterpress printing, principally in large **display** sizes.

Different forms of wood letter were widely used in the nineteenth century, particularly for the flamboyant playbills and posters that have since become collectors' items. Appropriately, their design often reflects this huckster purpose as well as the medium in which they were cut. Some of the letterforms are of such ingenuity and force (and sometimes of such surpassing vulgarity) that they remain popular today, translated into press type, film, and computer fonts. Original wood letter may still be found in the print shops of rural newspapers, though the market has been largely preempted by antique dealers and interior decorators. RE

Word block. *See* **Break block.**

Word division. *See* **Hyphenation.**

Word space. The space between words. In **unjustified** setting, word spaces should be fixed at a given value, expressed in **units.** In **justified** setting the word spacing will vary from line to line in order to fill the **type measure.** The "correct" amount will vary with typeface, size, measure, **leading,** etc. and is usually determined by the typesetter. If spacing is to be specified, it should be in units of the **em,** with 5 units of the 18-unit em being one common standard. This 5-unit space falls between the **thick space** (6 units) and **middle space** (4.5 units). A **thin space** would be 3.6 units and is the smallest "normal" space used with the instruction "close word spacing." Minimum and maximum amounts will vary as well. *See also* **Unit.**

Type justified to a very short measure is bound to produce unsightly rivers snaking down the paragraphs, as may be seen from the columns of almost every newspaper. There is little the compositor can do to improve this situation unless the client decides to reduce the type size or increase the measure. Adding noticeable letterspacing, except in display type, is unacceptable in good bookwork, though frequently resorted to in newspapers. To achieve even spacing in justified setting, the ideal number of characters per line is about 72, though reasonable spacing is often achievable with 60, depending to some extent on the average length of the words in the manuscript. A treatise peppered with hybrid Greco-Latin jargon is likely to set less tightly than a Harlequin romance. RE*

Typesetters prefer that designers not specify the word spacing to be used on a job, aside from general instructions like "thin spacebands" or "tight word spacing." There are a host of reasons for this, centering on their belief that good composition is not a matter of "magic numbers" (which, after all, could be programmed into a computer by anyone). Whether these assumptions are warranted is another question. CME

Word spacing. *See* **Word space.**

Wrap. To indent lines of type on either side or both sides to follow a contour. In magazine layout, type is often wrapped or formed around the shape of an illustration. In bookwork the term is used with particular reference to **drop caps,** when the abutting text lines are made to follow the profile of the initial rather than align vertically (follow a **box indent**). *See also* **Runaround.**

Wrong font (WF). Letters or words that have been miskeyed during typesetting and appear in the wrong typeface, style, or size. *See also* **Appendix 8.**

WSR. Abbreviation for *white space reduction. See also* **Minus letterspacing.**

Wyn, wynn (P, p). Old English character for w; also called *wen.*

WYSIWYG. Acronym for *what you see is what you get,* pronounced "wizzywig." A term used to describe a computer's ability to show the image on screen as it will print.

The screen representation is not exactly what will appear on the print-out, as will be attested by anyone who has tried to kern display type on the screen. The primary feature and limitation of WYSIWYG is that it is interactive, and some kinds of manipulation are hard to do. Also known as "what you see is all you get." CME

x-height. The height of the **lowercase** alphabet, discounting the **ascenders** and **descenders.** The letter x defines this measurement because of its usually horizontal **serifs** top and bottom. *See also* **Appendix 1.**

Yen symbol (¥). Character representing the unit of Japanese currency.

Yogh (ȝ, ȝ). Character used in Middle English to represent sound of y or gh, as in *light.*

Zinco (from Zincograph). (1) An engraved or etched zinc plate. (2) An illustration printed from an engraved zinc plate.

Zipatone. A trade name, sometimes used generically, for a wide range of patterns printed on self-adhesive transparent film and used in the preparation of **line art,** particularly to create shading effects or to distinguish various elements in maps and diagrams.

Appendixes

Rounded characters, such as o, e, and s, extend slightly above and below the x-height to maintain a visual rather than a mathematical alignment. In many typefaces the height of the ascenders is greater than the height of the capital letters.

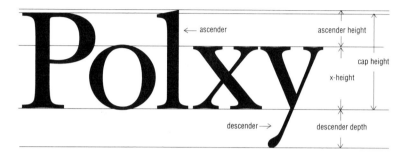

The counters of a letter, marked **c** in the figure below, are the white spaces enclosed or defined by the strokes of the letter.

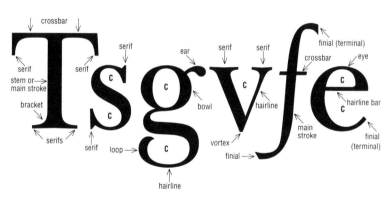

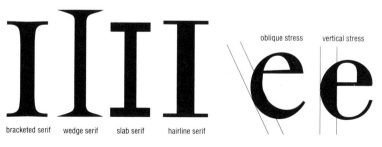

Relative Size

The figure below shows the variation in size from one typeface to another. Both examples are set in 49-point, but the Galliard on the left has a noticeably larger x-height than the Adobe Garamond on the right. The first two letters in each series are lowercase, the second two are small caps, followed by old-style figures. In most typefaces the small caps are taller than the x-height of the lowercase. The old-style figures are usually the same as the x-height of the lowercase or slightly shorter.

xaXA91   xaXA91

Another variable from one typeface to another is the relationship of capital letters to lining (modern) figures. Shown below, the Galliard figures are cap high, whereas the Adobe Garamond figures are smaller than the capitals.

X1234 X1234

Ligatures

The most common ligatures, fi and fl, are available in most text faces. Ligatures are sometimes called *tied letters*, though some reserve this term for combinations like the ct, which contains an obvious tie stroke.

ff fi fl ffi ffl ct

Swash Letters

Swash initials

Swash terminals

APR rtanz

Front matter (preliminaries, prelims), usually numbered with lowercase roman numerals

i	Bastard title (false title, first half title) or series title
ii	Ad card, series title, series list, frontispiece, or blank
iii	Title page (may be a double-page spread)
iv	Copyright page (includes copyright notice, printing history, country where printed, ISBN, CIP data [or notice that this appears on the last page of the book]; may also include publisher's address, copublishing information, acknowledgments, credits, permissions, dedication, epigraph, or colophon)
v	Dedication or epigraph (or table of contents)
vi	Blank
vii	(Table of) Contents opening page
recto or verso	(List of) Illustrations opening page
recto or verso	(List of) Tables opening page (may be run in after list of illustrations or start a new page)
recto	Foreword opening page
recto	Preface opening page
recto	Acknowledgments opening page (may be run in after preface or start a new page)
recto or verso	Other front matter (list of contributors, list of abbreviations, chronology, translator's note, note on orthography, etc.)
recto	Introduction opening page (if not part of main text)

Text, numbered with arabic numerals that continue through the back matter

1	Half title (second half title) or part title; the half title may be numbered as part of the front matter and is usually omitted if there is a part title

| 2 | Blank |
| 3 | Chapter 1 (opening page of main text); if the half title is omitted or numbered with the front matter and there is no part title, text may begin on page 1 |

Back matter (end matter), usually opening recto unless there are space constraints

recto	Appendix opening page; if there are several appendixes, each one usually starts on a new page, recto or verso
recto or verso	Notes opening page
recto or verso	Glossary opening page
recto or verso	Bibliography opening page
recto or verso	Index opening page
last verso or recto	Colophon and CIP data (if not on copyright page)

There have been many attempts to classify the hundreds of typefaces that exist according to common attributes such as stress, stroke weight, and shape of serifs. The most widely adopted method is historical, but there is continuing controversy over the definition of historical periods and how they should be described. What follows is a summary and by no means authoritative. Dates are approximate and refer to the creation of the earliest examples within a group. Typefaces designed at a later period are included in the historical category whose attributes they share.

Black letter or **gothic** (1465). Example: 𝔊𝔬𝔲𝔡𝔶 𝔗𝔢𝔵𝔱
    Characteristics: derived from the broad-nib pen style of medieval manuscripts written before the establishment of the humanist Caroline minuscule hand that more closely resembles our contemporary alphabet; spiky vertical stress, typically without any curved strokes; diamond-shaped serifs; often includes double strokes of contrasting weight separated by a white line. Also called Old English.

Venetian (1470). Example: Centaur
    Characteristics: derived from Caroline minuscule; little contrast between thick and thin strokes; diagonal stress, the diameters of the rounded letters slanting backward; steeply sloped serifs; lowercase e has a diagonal bar. Venetian is often included as a subcategory within old face.

Old face (1495). Example: Bembo
    Characteristics that differ from Venetian: slightly more contrast in stroke weight; narrower in the lowercase; serifs less slablike, with a rounder angle; capitals tend to be shorter than lowercase ascenders; the lowercase e has a horizontal bar.

Transitional (1700). Example: Bell
    Characteristics: more contrast in stroke weight than found in old face; stress closer to upright; x-height relatively greater; serifs not triangular, but roundly bracketed and tending toward the horizontal.

Modern (1800). Example: Bodoni

Characteristics: marked contrast in stroke weight; vertical stress; unbracketed, thin, horizontal serifs; capital letters same height as lowercase ascenders; rounded letters circular in outline.

Egyptian or **slab serif** (1820). Example: **Rockwell**

Characteristics: almost no difference in stroke weight; emphatic horizontal serifs with no bracketing.

Grotesque (1820). Example: Akzidenz Grotesk Condensed

Characteristics: similar to Egyptian in the subtle contrast of stroke, but without serifs. Though sans serif in the literal sense of "without serifs," distinguishable from that group by a less geometric construction. Some typefaces of this group are called gothic, perhaps because their heavy color recalls black letter.

Clarendon (1845). Example: Century Schoolbook

Characteristics: similar to Egyptian in verticality, but with strong, bracketed serifs and some variation of stroke weight; emphatic in overall effect.

Sans serif (1916). Example: Futura

Characteristics: as the name implies, without serifs; geometric in construction and virtually without modeling or change in line weight.

Other, catchall categories are:

Twentieth-century romans. Text faces not conforming to any of the groups above. Examples: Joanna, Zapf International, Cartier.

Script and brush. Examples: *Shelley Allegro, Zapf Chancery, Mistral, Balloon. Characteristics: directly imitative of handwriting or pen, engraved, or brush lettering.*

Decorative and display. Examples: Broadway, Neuland, Umbra, Peignot. Characteristics: strongly individualistic and defying category; primarily for use in large sizes as display headings or initials. Within this group might be included outline, inline, and shadow typefaces as well as the anciently derived uncials such as Libra and Zapf's classical Michelangelo, a display face by intention, made up of capitals only.

Once a manuscript is designed, clear and complete typesetting specifications must be written. When typesetters translate a design into typesetting formats, they usually prefer to work from a complete separate sheet of written specifications, not just those marked on the manuscript or layouts. Formats are established not only for the text but for all repeating elements such as chapter titles, subheads, and extracts.

To simplify the work of the designer, the editor should code the different elements of the manuscript as part of the editing. Every publishing house has its own set of codes, usually simple designations such as PT for part title, PN for part number, CT for chapter title, CN for chapter number, A or 1 for the primary order of subhead, B or 2 for the second order, and so on. Once the designer has given the specification for the first occurrence of such an element, such as an A-head, it is unnecessary to repeat it every time that element reappears. Only special or unusual cases need to be marked.

In addition to preparing standard specifications, it is important that the designer clearly inform the typesetter of all deviations from the standard. Some of the specifications should be marked directly on the manuscript. Although specifications should be succinct, it is sometimes better to explain in complete sentences how to typeset some of the more complicated parts of a manuscript rather than to rely on abbreviated instructions that may be misunderstood. Whenever possible, a visual example of how something is to look should be supplied. It is especially useful to supply designs and complete markups for pages in the front matter that are unique, such as the half title, title page, copyright page, dedication, and epigraph. Many designers prefer to show how display headings are to be broken, and in this case line breaks should be marked on the manuscript.

In the United States a book's trim size is expressed in inches, with the width preceding the depth. Measurements for the margins are usually given in inches, although some designers prefer to use picas and points. All other measurements for the type page are given in picas and points. Measurement from one element to another is generally given from base to base (b/b). For example, a specification might read "24 pts b/b from running head to first text line." Using the base-to-base measure removes any confusion about exactly how much space is wanted. The base of a line of type is measured from the bottom of the x-height, not from the bottom of the descender.

Some typesetters like to have the overall type page expressed. The overall type page is measured from the top of the first element to the base of the last (that is, from the top of the running head to the base of the last text line) and from the leftmost element to the rightmost. This measurement does not include special elements such as drop folios, which might appear only on chapter openings and are therefore not part of the regular type page.

The top margin is measured from the top trim (that is, the top edge of the paper) to the top of the topmost element on the page (that is, the top of the cap of the running head or the first line of text). Establishing the gutter margin is straightforward unless the text page is of variable width, as in a book of poetry. In that case it is necessary to establish a point of reference, such as the folio or some other constant element on the page.

Because of the profusion of typefaces with the same name, it is essential to specify not only the name of the face but also the foundry and the method of setting. For example, Linotron 202 Garamond no. 3 is considerably different in look, size, and weight from Adobe (PostScript) Garamond or Monotype Garamond.

Lining or old-style figures should be specified. If both styles are used in the book, the specifications should tell how each is used (for example, "old-style figures in text; lining figures in table headings and table bodies").

Paragraph indentions are specified either in points or in ems. If various sizes of type are used in the text, specifying the exact number of points of a particular size em (for example, 10-point em) will ensure that elements will align with each other. A 10-point em is quite different from a 14-point em.

In specifying type, size is expressed before leading. Thus the specification 10/14 × 27 Bodoni means 10-point type set with 4 points leading, to be set justified 27 picas wide. Ragged right would be specified as 10/14 ragged right × 27 picas maximum. Some designers prefer to specify a minimum measure also to keep the right edge from being excessively ragged. In ragged-right setting one must state whether hyphenation is acceptable.

In addition to specifying the type style for the main text, give the same information for extract material. Specify indentions, if any, and the space above and below the extract. If there are successive extracts, specify the space between them. For poetry extracts, it is necessary to say whether the lines are to be indented, and how much. Many designers like to center the longest line of poetry on the text measure and align all the other lines with the longest, which gives the poetry extract the appearance of being centered as a block of text. Other designers prefer to establish a fixed indention. Stanza breaks must also be marked, and specifications are needed for lines of poetry that are too long to fit the text measure so that turnovers are necessary. If sources for the extracts are included, their position – run in with the text or set on a new line – must be specified.

The designer needs to specify the position and style of running heads and folios, as well as the space between running head and folio if they are on the same line, and to note the space between the running head and the first text line. Specifications for drop folios should give the space from the base of the last text line to the base of the folio.

Display type is the most complex element to specify. One sure method of indicating placement is to give the text line it aligns with (for example, "chapter number aligns with 4th text line"). If display type is letterspaced, it is useful to show an example in the layout. Because of the variations in type design and typesetting methods and measurements, it is extremely difficult for the typesetter to determine how much space is wanted without seeing an example.

Since many publishers prefer to have facing pages align at the bottom, the designer should try to make spacing above and below sub-heads in the text add up to full text lines. Irregular amounts of space around subheads will result in uneven facing pages.

Following are the typographic specifications for the *Glossary of Typesetting Terms:*

GLOSSARY OF TYPESETTING TERMS
Specifications for typesetting
Book trim size: 5½″ × 8½″
Margins: 3 picas top, 4 picas gutter
Typeface: Adobe Garamond
    See also below for Appendix 3 examples
Figures: old style
Paragraph indent: 1 11 pt em
Main text: 11/15 × 24, 34 text lines. Card pages to align if necessary,
    otherwise OK one line short facing pages
    Use all f ligatures
    Use spaced en dash instead of up close em dash
Extract: none
Running foot & folio:
    Running foot in 10 pt italic Clc
    Folios in 10 pt old-style roman figures within word-spaced brackets
        Brackets indent 1 pica from outside margins
        Running foot is en space from folio
        27 pts b/b from last text line to running foot
        Do not bounce on short pages
Half title [i]: see layout
    14 pt even small caps letterspaced as on layout (space shown is 50%)
    32 pt italic Clc with swash cap; note position of word *Typesetting*
Series page [ii]: see layout
    Logo is on 5th text line
    30 pts b/b first entry
    11/15 as broken on layout
    Titles in roman, authors in italic indent 1 pica
    30 pts b/b between entries
    Two columns, second column indent 13 picas
Title page [iii]: supplied on disk by designer
Copyright [iv]: see layout
    9/15 ragged right. Two columns, begin on 3d text line
    1 pica column space

First column is ragged right 13 picas maximum
Second column as broken
Contents [v]: see layout
    Heading 11 pt even small caps letterspaced 50%
        Flush left on 1st line
    30 pts b/b 1st line of text
    11/22.5 Clc flush left
    Appendix titles indent 1 pica
    Folios separated from titles by comma and word space
Introduction [vii]: see layout
    Heading 11 pt even small caps letterspaced 50%
        Flush left on 1st line
    30 pts b/b 1st line of text, which begins flush left Clc
    Text: 11/15 × 24. Folio will show on opening page
Glossary part title: arabic page 1
    30 pt italic with swash initial, flush left on 3d text line
Glossary text begins page 3
Glossary
    Headnote subhead 11 pt even small caps letterspaced 50%, flush left on 1st line
    Headnote: 11/15 as broken
    Section initials: 60 pts base of text to base of initial
        48 pt italic initial or swash initial flush left
            Use swashes only for the following:
            A F J K L M N Q R T Z
    21 14 pt periods to fill the measure, center on cap height of initial letter
    30 pts base of initial to base of 1st text line
    There must be at least two text lines below initial; otherwise end page short and
        begin new page with initial on 3d text line
    Entries begin flush left in 11/15 semibold followed by period and word space. Text
        in 11/15 × 24. Turnovers indent 1 em
    See and *See also* in italic
    Cross-references in semibold
    No extra space between entries
    Special instructions
        Manuscript pages 29 and 38 use 10/15 Gill Sans for example
        Manuscript page 31 use 8/10 Gill Sans for example
        Manuscript page 72 use 24 pt Arabesque ornament number 1 keystroke shift
            W
        Manuscript page 79 set ½ pt box 6 × 9 picas
            Position flush right
            Text ragged right 17 picas to run around box
Appendix part title: recto
    30 pt italic Clc with swash cap, flush left on 3d line

Appendixes begin new page verso or recto except Appendix 1, which begins new
recto

Appendix number 11 pt even small caps and old-style figures letterspaced 50%
flush left on 1st text line

Appendix title 14 pt italic Clc on line with number
En space between number and title

Begin text on 3d text line

Appendix 1 is artwork

Appendix 2:
Two columns, ragged right 24 picas maximum
Turnovers indent 1 em, 1 pica ditch

Appendix 3 subheads in semibold flush left
Examples to be set in different typefaces as indicated
11/15 × 24 indent 1 em
Goudy Text, Centaur, Bembo, Bell, Bodoni Book, Serifa (or Rockwell),
Grotesque Number 2, Century Schoolbook, Futura, Joanna, Zapf
International, Cartier, Shelley (or Kunstler), Broadway, Peignot

Appendix 4 is normal text

Appendix 5:
Set G & S style sheets 9/12 × 24
A-heads: 11/48 semibold Clc flush left
24 pts b/b below A-heads
Section heads: 9/12 even small caps letterspaced 50%
24 pts b/b above, no extra space below
Align with text
Open boxes and old-style figures for entries will hang 1 word space from text
Footnotes: 8/11 × 24

Appendix 6:
A-heads 11/15 semibold Clc flush left
37.5 pts b/b above, 22.5 pts b/b below
Begin text flush left Clc
B-heads 11 pt italic Clc indent 1 em
30 pts b/b above, no extra space below
Begin text Clc indent 1 em
Use table examples in manuscript as art

Appendix 7:
Ragged right line for line, two columns, 1 pica ditch

Appendix 8:
Artwork
Set footnote 8/11 × 24

Bibliography: recto
Heading 11 pt even small caps letterspaced 50% flush left on 1st line
30 pts b/b first entry

11/15 ragged right 24 picas maximum. Begin entries flush left
    Turnovers indent 1 em. Do not hyphenate
Contributors: recto
    Heading 11 pt even small caps letterspaced 50% flush left on 1st line
    30 pts b/b first entry
    9/12 × 24. Begin entries flush left
    Turnovers indent 1 em
    18 pts b/b between entries

The most comprehensive and generally accepted reference for house style in its broadest sense continues to be *The Chicago Manual of Style*. Prepared with remarkable thoroughness, it provides a standard to which publishers may refer in-house and freelance staff, authors, and typesetters to ensure consistency of presentation. The editors of the *Chicago Manual* have always stressed that it is intended as a reference tool rather than a set of inviolable commandments. Used in this way, it provides a model by which difference as well as agreement may be defined.

The *Chicago Manual's* thorough coverage of copy preparation makes it particularly valuable to editors. It also examines the technical aspects of bookmaking. There is an evident need among those who commission or supply typesetting, however, for a much briefer document that deals specifically with typography. It is this aspect of house style, sometimes referred to as *composition rules,* that is the subject of this appendix.

During the compilation of this glossary we have been very much aware of the lack of agreement about the meaning of even some of the most commonly used typographic terms. A style sheet, by identifying the elements of typographic style and by defining terms, attempts to avoid the confusion and misunderstanding that so often frustrate both client and typesetter. Surprisingly few publishers provide this guidance, either out of ignorance or because of a false assumption that consensus exists. This omission does not invariably lead to disaster. If a publisher works regularly with the same supplier, the quality of work is likely to be correspondingly consistent. But many publishers, particularly those that are required to bid out each job to a number of vendors, never develop such a relationship. In this situation a style sheet is essential.

The fact that typesetters work with many clients, whose demands may be conflicting or unspecific, has prompted them to establish their own standards, which they will apply unless instructed otherwise. Many of those who specialize in bookwork have produced excellent documents of this sort. For example, G & S Typesetters of Austin, Texas, routinely supplies copies of its style sheets to customers at the inception of a project, not with the intention of imposing standards but as a way of establishing the individual requirements of the publisher. With the kind permission of G & S, these documents are here reproduced in the same spirit in which they were formulated – not as rules set in stone, but as a checklist of items that need to be considered.

G & S House Style

This sheet reflects the house standards for typesetting at G & S Typesetters. We have modified these instructions where we are aware that your preferences differ from our usual practice. Please look over these instructions and, where you wish us to apply different standards from those specified, indicate by checking the box at the left and writing complete instructions (keyed to the item number) on the blanks at the end of this form or on additional pages. Please return this sheet with one set of approved sample pages.

EQUIPMENT

☐ 1. Our normal output device for phototype is the Linotron 202, which can set type from 4½ point to 72 point, in half-point increments. For 48-point type and above, we prefer to set type on the Phototypositor (provided we have the typeface in our typositor library) because of superior quality.

STYLES OF PUNCTUATION

☐ 2. Following the *Chicago Manual of Style*, we set punctuation in the same font as the word, letter, or character immediately preceding; e.g., following an italic word, a comma or question mark will be italic; following a bold word, a colon or dash will be bold. However, we make exceptions to this rule in scientific and computer texts, where the font may have a specific meaning. The rule does not apply to paired marks of punctuation: parentheses, brackets, and quotation marks are usually set in roman, except when parentheses or quotation marks are part of an italic title.

☐ 3. If the author's computer files are furnished with a manuscript and we are able to translate word-processing codes into typesetting font changes, we will leave the style of punctuation as it was coded by the author or editor, in order to minimize the amount of handwork required to process the files.

☐ 4. Inconsistent placement of punctuation with regard to quotation marks or superior footnote reference figures will be normalized to standard American usage: periods and commas will always be placed inside single or double quotes; colons and semicolons will always be placed outside single or double quotes; question marks and exclamation points will be placed inside or outside depending on whether they are part of the quotation or part of the surrounding sentence; superior reference figures should always follow any mark of punctuation.

WORD DIVISION

☐ 5. Our authorities for hyphenation are *Webster's Ninth New Collegiate Dictionary*, *Webster's Third New International Dictionary*, and *The Chicago Manual of Style*.

☐ 6. Maximum permissible number of consecutive end-of-line hyphens is 3.

☐ 7. We observe the following rules for hyphenating words:
- Never break a single-syllable word.
- Never break anywhere except at a legitimate hyphenation point that can be demonstrated by reference to the above-mentioned authorities.
- Avoid hyphenating the last word of a paragraph.
- Avoid hyphenating the second word in a hyphenated compound.
- Avoid hyphenating an acronym.
- Avoid hyphenating after a single letter at the beginning of a word or before 2 letters or fewer at the end of a word.
- Avoid breaking before any of the following final syllables: -ble, -tle, -gle, -cle, -fle, -ple, -dle, -zle, or any of these combinations followed by an *s* or *d*.
- Avoid hyphenating before a single-letter syllable in the middle of a word.
- Break closed compound words between the two parts of the compound.
- Avoid breaking numerals.
- Avoid hyphenating any part of a person's name.

☐ 8. Widows: The last line of a paragraph must contain at least 5 characters plus the ending punctuation, should not consist entirely of part of a hyphenated word, and should always be wider than the paragraph indent.

☐ 9. We also observe the following rules for line endings that do not involve hyphenation:
- The same word should not begin or end more than two successive lines (word block).
- Avoid breaking between a person's first and middle names, whether spelled out or initials.
- In a run-in numbered or lettered list, do not separate the number or letter from the copy that follows it.
- Avoid breaking between an abbreviation and a number, such as "p. 100," or between a number and its unit of measure (e.g., 4 cm).
- If possible, avoid breaking a line before an em dash or a three-dot ellipsis.

- The last line of a paragraph should be justified if it is less than 1 em short of full measure.

☐ 10. Leading: The leading given in our specs expresses the distance from the previous baseline to the baseline of the current line.

☐ 11. Measure: Our specifications always give the overall measure and then indicate any indents on left or right that reduce that measure.

☐ 12. Paragraph indents: If not specified, we assume a paragraph indent of one em. For any text elements that are set flush left with the main text, such as unindented extracts or footnotes, we will use the same size paragraph indent as the text. If no general instructions are given, we will follow the way the manuscript is typed for paragraph indents on extract copy.

☐ 13. Units: Our unit is $\frac{1}{18}$ em.

☐ 14. Figures: If not specified, we assume lining figures are called for. Superior reference figures and fractions will always be set with lining figures, even if old-style figures are specified for text. Within the body of tables we always use lining figures, unless old-style figures are specifically requested. In titles and subheads that are set in all caps, we will always use lining figures; in titles and subheads set entirely in small caps, we will always use old-style figures; in titles and subheads set in caps and small caps, we will use whichever style of figures is used in the text.

☐ 15. Ragged right setting: Normal procedure is to allow hyphens in ragged right copy. We try to maintain a maximum variation of line length of 2 picas wherever possible.

☐ 16. Justified setting: The maximum allowable space between words is 10 units. We never increase or decrease letterspacing to justify lines.

☐ 17. Acronyms: All acronyms are set closed up, unless otherwise specified.

☐ 18. Dashes: En dashes are used to indicate a range of numbers, times, dates, or pages: 8–10 A.M., pp. 144–78. Em dashes are used to indicate a break or interruption in a sentence.

☐ 19. Rule weights: The finest rule we can set on the 202 is approximately $\frac{1}{3}$ point; this is the rule we set when a hairline or quarter-point rule is requested. The next heavier rule is approximately a $\frac{2}{3}$-point rule; when a half-point rule is specified, this is the rule we use. From 1 point and above, we can set any weight rule in half-point increments.

☐ 20. Fractions: Three styles of fractions are possible within text: case fractions ($\frac{1}{2}$), piece or autofractions (½), and full-size or shilling fractions (1/2). If the style is not specified, in technical books we will use piece fractions and in nontechnical books we will use full-size fractions.

☐ 21. Abbreviations and acronyms: All abbreviations and acronyms (including A.M./P.M. and B.C./A.D.) will be set caps (not small caps) unless we have different instructions.

☐ 22. Display initials: Without specific instructions, we will indent text lines an equal amount for a dropped initial cap (block indent style). Unless instructed to do otherwise, we will delete any quotation marks that precede a display initial, either stick-up or dropped.

☐ 23. Numbered and lettered lists: Unless instructed otherwise, we will allow for two-digit item numbers only in lists with 10 or more items. Because letters have varying widths, lettered lists will align on the left, not on the periods; the following text in each entry will still have the same alignment.

☐ 24. Poetry: Poetry that is to be centered on its longest line will not appear centered in galleys. (See Paging Instructions for further information about centering poetry.) Omission of lines of poetry will be indicated by a line of em-spaced periods as long as the preceding line.

☐ 25. Subheads: If not specified or indicated on the manuscript, we assume the first text line following a title or subhead should be set flush left. If subheads make more than one line, they will be broken for sense, to approximately equal lengths, always avoiding a line consisting of a single word. When two subheads stack, we will if possible use the higher-level head's spacing between them (and if necessary alter the space above the first head to make the pages align). If a subhead precedes a separate text element such as an extract or list, we will use the spacing as specified for the subhead.

☐ 26. Notes: We have no house style for either footnotes or endnotes; we rely completely on the design in all cases. If note numbers are superior, they are followed by two units of space before the note text; full-size note numbers are followed by an en space. Superior numbers are always set in the same size as the rest of the text element where they occur (i.e., they will be different sizes in the text and in the notes.)

☐ 27. Bibliographies: There are two styles for spacing around colons in journal references: with 2 units on either side, or closed up to the preceding word with a spaceband following. If specifications are not given, we will follow the spacing shown in the manuscript if consistent; if the manuscript is inconsistent, our house style is to use 2 units when the colon is between two numbers (e.g., 1990:3) and closed up to a preceding word or parenthesis (e.g., (August 1987): 45).

☐ 28. Indexes: Normally indexes are set ragged right, in two columns with a 1-pica gutter. If they start new lines, subentries indent 1 em; subsubentries indent 2 ems, and so forth; all turnovers will indent 1 em more than the lowest level of subentry. The main entry is followed by a comma before the first page number. En dashes are always used for inclusive pages.

☐ 29. Running heads: We have no house style that tells us what to use for running head copy; we always rely completely on instructions from the press. If drop folios are not mentioned in the specifications, they will not be set or pasted up.

30. Marginal flags: All numbered tables, figures, boxed text, or other illustrative material will be typeset separately from the main text, and a small notation will be typeset in the right margin at the end of the text paragraph in which the illustration was first mentioned. If requested, we will also flag cross references in the same way. Page numbers for cross references will be typeset as zeroes, unless other instructions are furnished.

31. Queries: All queries will be addressed to the editor and written on gummed slips that will be transferred to the repro before proofs are pulled, so that they appear on all proof sets. The only alterations we will make to the copy without querying are corrections of simple misspellings.

TABLES

32. The following is our house style for table typesetting:
    - Ideally, a table should appear as balanced as possible, with equal space between all columns in the body of the table (not necessarily between all column heads), disregarding occasional percent signs, plus or minus signs, superscripts, or extra figures that do not cause the visual center of the column to change.
    - The maximum allowable space between columns is 4 picas; 2½ or 3 picas is considered ideal; 1 pica is the minimum.
    - Column heads may be broken and words in column heads hyphenated as necessary to maintain optimum spacing in the body of the table.
    - All column heads should base align.
    - Type should be indented left and right an amount approximately equal to half the space between columns, with rules extending outside table text.
    - Generally, columns of words are aligned left and centered as a block under the column head, and columns of numbers are aligned on the decimal and centered under the head.
    - Tables will be set upright whenever possible. If we can maintain intercolumn spacing between 1 and 4 picas, we will set a table to the exact text width; but if this procedure would cause us to exceed the 4-pica maximum column spacing, we will hold intercolumn spacing to approximately 2½ picas and set the table as narrow as is required, with a minimum width of 16 picas. If, however, a small table has very long notes, we will increase the intercolumn spacing to 4 picas. An upright table may also exceed the text measure by up to 3 picas if necessary.
    - If a table can not be set upright and maintain at least 1 pica between columns, it will be set as a broadside table. Ideally it would be set to a width equal to the type page depth, provided intercolumn spacing can be kept between 1 and 4 picas; otherwise, column spacing will be held to 2½ picas and the table set as wide as is required.
    - A table too wide to fit broadside on the page may be set for pasteup across facing pages.

☐ Table titles, notes, and rules are set to the full measure of the table, the rules extending beyond the body of the table on each side by an amount equal to half the intercolumn spacing. If the main text of the book is justified, the notes to tables will be justified. Source should precede notes. Notes to broadside tables will be set to the full width of the table unless specified otherwise, or unless there is a large number of very brief notes, in which case they will be double-columned.

☐ Turnovers in the stub will be indented 1 em unless extra lead is indicated between items. If there is only one column with a turnover, rows usually align horizontally with the last turnover line; if there are two or more columns with turnovers, the rows usually top align.

☐ Letters used as superscripts for note references in tables will be set italic.

☐ Unless vertical rules are specifically requested, vertical lines in manuscript will be ignored.

Please list the item numbers which deviate from your desired style and state your preference:

_____   _____
_____   _____
_____   _____

G & S Paging and Camera Instructions

This sheet reflects the house standards for page makeup and camera work at G & S Typesetters. We have modified these instructions where we are aware that your preferences differ from our usual practice. Please look over these instructions and, where you wish us to apply different standards from those specified, indicate by checking the box at the left and writing complete instructions (keyed to the item number) on the blanks at the end of this form or on additional pages. Any item that is highlighted is a point on which we will need further specific instructions from you before we can proceed to pages. Please return this sheet with one set of approved sample pages.

PAGINATION
Arabic page numbering (page 1) begins with: ☐ Second half title   ☐ First part title   ☐ Introduction   ☐ First chapter title   ☐ _____
(Usual practice of the press, if known:_____)

GENERAL PAGE MAKEUP
☐ 1. Depth of page and allowable variation: see spec sheet. In default of other instructions, we will allow pages to go up to 1 line short or long as necessary, except if running feet or drop folios occur on all pages, in which case we will only go short. (Usual practice of the press, if known:_____
_____)

☐ 2. Chapter 1 will always begin on a recto page, even if arabic page numbering begins with a preceding section; an epilogue, afterword, or other unnumbered chapter that follows the last numbered chapter will always begin on a recto. Unless specified otherwise, all front- and backmatter titles and all chapter openings that follow a part title will be recto. If not otherwise specified, we assume other chapters can open recto or verso. Multiple appendixes or indexes will follow the specs for chapter openings (either recto/verso or recto only) after the first appendix or index, which opens recto only.

☐ 3. Base-to-base measurements are unequivocal and therefore are preferred for indicating leading. The term "sink," on the other hand, has acquired several meanings, and is often ambiguous. When we encounter a vertical measurement given in terms of "sinkage," we will try to determine in what sense the term is used; if there is room for doubt, we will interpret the term "sink" to mean a measurement from the top of the running heads (i.e., the top of the type area) to the cap height of the line in question. Unless specified otherwise, we assume the top margin is a measurement to the top of the running heads. We define *type page depth* to mean the distance from the top of the running heads to the base of the last line of text (or from the top of the first text line to the base of the running feet); we use *text page depth* to mean the distance from the top of the first text line to the base of the last text line. Tick marks will show on all pages without running head and folio. If the position of the first text line in front- or backmatter is not specified, we assume it sits on the same baseline as a normal chapter opening.

☐ 4. Space will be varied above subheads and footnotes, and equally above and below extracts, equations, poetry, lists, etc., if necessary to make facing pages align or to avoid bad page breaks. Carding or feathering will not be used.

☐ 5. The number of lines on a chapter opening page will be as specified by the design unless this would produce a bad page break, in which case the sink to the first text line may be altered to vary the number of lines by up to 2.

☐ 6. Minimum number of lines of text below a stand-alone subhead at the bottom of a page is 2. A run-in subhead must be followed by at least one full line of text.

☐ 7. If a subhead is not the first element on a page, minimum number of lines of text above is 2.

☐ 8. Subheads that fall at the top of a page will be sunk the minimum amount so that the first text line following aligns with a text line on a normal page grid.

☐ 9. A run-in chapter title must be preceded by at least 3 lines of text and followed by at least 3 lines, unless the title is especially large or has a great deal of extra leading above or below, in which case we will try for at least 4 lines before and after. If the title falls at the top of a new page, the head will be sunk the full amount indicated in the specs.

☐ 10. In extreme circumstances, a text page may be up to 4 lines short if a subhead falls at the top of the next page. We will do this only as the last resort.

☐ 11. If a separate text element such as an extract, poem, list, etc., is not the first or last element on a page, minimum number of text lines above or below is 2.

☐ 12. Minimum number of lines of a separate text element that may fall at the top or bottom of a page is 2.

☐ 13. Minimum number of lines of text that may appear on a page with an illustration or table is 5.

☐ 14. Minimum number of lines on the last page of a chapter is 5.

☐ 15. A designated spacebreak should have at least 2 lines above if it falls near the top of a page or 2 lines below if it falls near the foot. If necessary, a spacebreak may fall between pages and the extra space will be left at the foot of the page. If the space absolutely can not be left at the foot of the page, no extra space will be left anywhere (the following page will always begin on the first text line as usual). Space breaks will never be allowed to fall between pages when there are footnotes.

☐ 16. First line of a paragraph may fall at the foot of a page; last line of a paragraph may not fall at the top of a page (or beneath an illustration at the top of a page), unless it is full measure. (For a two-column book, a short line is acceptable as the first line of the right-hand column if it precedes a displayed equation.) Last word on a page may be hyphenated. Last word of a paragraph will not be hyphenated unless unavoidable; last line of a paragraph may not be shorter than the paragraph indent unless unavoidable. We will try not to break a page within in-text table columns that total, or after a short line introducing an extract or equation. In the table of contents, we will never break between a part title and the following chapter title.

FOOTNOTES AND ENDNOTES

☐ 17. Unless specified otherwise, note numbers will align on the period if they are hung to the left of the text, but not if the notes are paragraph style. If note numbers are to align on the period, we will indent all single-digit notes within a given chapter an additional figure space, but will indent for alignment only those two-digit numbers that fall on the same page with three-digit numbers.

☐ 18. In default of other specifications, the space above footnotes will be a minimum of about 18 points base to base up to a maximum of about twice the text leading.

☐ 19. Footnotes and endnotes will not be double columned.

☐ 20. Long footnotes may be broken from one page to the next. If possible, at least 2 lines of the footnote will appear on each page and the note will be broken in the middle of a sentence. A 5-pica hairline rule will be inserted above the continued part of the note, in the same position as the baseline of the first line of a regular footnote and with no extra space below.

☐ 21. Footnotes that appear on the last, short page of a chapter will follow the text by standard leading (i.e., they will not drop to the bottom of the page). On other short pages within a chapter (such as those that precede a subhead at the top of the next page), footnotes will remain in their normal position at the bottom of the text page.

☐ 22. If endnotes have extra leading between notes, facing pages in the note section will not be able to align at the bottom.

POETRY

☐ 23. If poetry is to be cut to center (i.e., centered on its longest line), all poems that fall on the same page of text will hold the same left alignment; poetry that continues to a new page will be centered on the longest line that falls on the new page.

☐ 24. If poetry must be divided other than between stanzas, it will not be divided after the first line or before the last line of a stanza or, if possible, between rhymed lines.

BACKMATTER/INDEX

☐ 25. If absolutely necessary, the depth of facing index or other backmatter pages may vary by one line.

☐ 26. If necessary, widows are permitted at the top of columns in backmatter and index, provided they are at least three-fourths full line length.

☐ 27. In a bibliography, when a repeated entry begins a verso page, we will leave the 3-em dash where shown on the manuscript; if the editor marks the page proofs to change the dash back to the author's name, the change will *not* be counted as an editorial alteration.

☐ 28. Continued lines will be used for index entries that break from a recto to a verso page. They will consist of the main entry (followed by an em dash and the subentry, if necessary) followed by "(*continued*)" or "(*cont.*)" depending on space available. If necessary, a continued line may take up 2 lines of the index.

☐ 29. Minimum number of lines on the last page of an index is 5 in each column; columns on this page may vary in length, by one line if possible or more if necessary, with the first column being longer.

☐ 30. Name index precedes subject index.

☐ 31. Space may vary between alphabet breaks when necessary.

☐ 32. A single line followed by a break between alphabet sections should not fall at the top of a column, nor should a single line following a break fall at the bottom of a column.

☐ 33. Index pages will be rerun as needed for corrections that require repaging. Patches will be set if type need not be run from column to column.

Running head list:　□ Furnished　□ To come

☐ 34. Running heads or running feet and folios that sit on the same baseline are typeset as a single piece of repro whenever possible, for ease of pasteup and greater stability.

☐ 35. Running heads and folios will be left off any page that contains an illustration but no text, and tick marks will be added.

☐ 36. Folios or running feet that are to be positioned at the foot of the page throughout the book will maintain the same position relative to the bottom trim, regardless of the length of the page (i.e., they will not bounce).

☐ 37. If, in the main body of the text, the recto running heads use different copy and are a different style from the verso (e.g., verso: book title in small caps; recto: chapter title in italic clc), for front- and backmatter heads that read the same recto and verso (e.g., Contents, Preface, Index), we will use the recto style for both pages.

☐ 38. Running heads will be retained on pages that begin with run-in chapter titles.

☐ 39. For text-related running heads (i.e., those that use a subhead or run-in chapter title for copy), we will use the subhead pertaining to the first text on a verso page for the verso running head, and the last new subhead on a recto page for the recto running head.

☐ 40. For self-indexing running heads in a section of endnotes ("Notes to Pages 67–83"), we will leave these off the first set of page proofs and expect the editor to write them in.

ILLUSTRATIONS (TABLES, FIGURES, MAPS, PHOTOS, BOXES)
This project will include (indicate number of pieces):
Tables _____ Figures _____ Maps _____ Other _____
"Floating" plate section, for press placement: _____ pages
Folioed (or blind folioed) plate section: _____ pages

☐ 41. Positioning: Illustrations will be positioned as close to their text citation as possible, preferably after that citation but before the next major subhead, at the top of the page when possible, or at the foot if necessary. If two illustrations fall on the same page with room for at least 5 lines of text, the text will preferably be placed at the bottom of the page. When illustrations of different depths occur on facing pages, where possible one will be placed at the top and the other at the bottom of their respective pages. An illustration will never be placed between text and a footnote. If possible, no illustration will be placed on the first or last page of a chapter, or facing the first or following the last page of a chapter; or immediately before a subhead; or within an extract, list, or other separate text element. An illustration will never be placed immediately after a subhead.

☐ 42. Upright illustrations may extend up to 1½ picas into the right and left margins. Illustrations narrower than text width will be positioned on the page (flush left or centered) according to the general style of display type throughout the design. A table that is too long to fit on one page will be placed on 2 facing pages, if possible. Notes to a table that continues to more than one page will be placed at the end of the table. Each page of a continued table will be headed with the table number and "(*continued*)" in the table title size and position.

☐ 43. An illustration more than 3 picas wider than text will run broadside, on a verso page if possible; it will be centered vertically and horizontally on the page. If a broadside illustration continues to a second page, we will leave the same gutter margin between both halves of the illustration. Notes to broadside tables will be the width of the table unless specified to double column, and will be placed at the end of the entire table.

☐ 44. Space allowances given for illustrations are interpreted to include actual illustration only; caption and extra space above or below are additional. If no specs are provided for space between an illustration or caption and text, we will default to approximately two text line spaces base-to-base, with a variation of about 6 points either way.

☐ 45. In 2-column texts, when two illustrations fall on the same spread we will try to avoid placing both in the gutter or both in the left or right columns. Adjacent illustrations should be the same depth when possible; space below may vary by up to 3 lines if necessary to keep text aligned in both columns. Three or more illustrations should preferably be grouped on the spread, not broken by a column of text; extra white space should be at the top and outside.

☐ 46. Boxed text: "Floating" boxed text that runs more than a page long will be broken into equal halves across the spread and the rules will extend across the gutter. If such a box runs more than 2 pages long, a continued line will be added to the recto page and another to the beginning of the following verso. If boxed text is an integral part of the text (not floating), it will break as necessary, with no continued lines.

CAMERA WORK

☐ 47. Unless otherwise specified, halftones and screen tints will be shot with a 133-line screen.

Please list the item numbers which deviate from your desired style and state your preference:

—— ————————————————————————————————
—— ————————————————————————————————
—— ————————————————————————————————

G & S Math and Technical House Style

This sheet reflects the house standards for typesetting and paging math and technical material at G & S Typesetters. We have modified these instructions where we are aware that your preferences differ from our usual practice. Please look over these instructions and, where you wish us to apply different standards from those specified, indicate by checking the box at the left and writing complete instructions (keyed to the item number) on the blanks at the end of this form or on additional pages. Please return this sheet with one set of approved sample pages.

AUTHORITIES

☐ 1. Our authorities in math and technical matters are *Mathematics into Type* and *Chicago Manual of Style.*

TYPESETTING

☐ 2. Variables: Letters used as mathematical variables (distinguished from abbreviations of 2 letters or more) are always set italic unless specifically marked to be roman. Numbers are always set roman. Both these rules hold true even within a text element (such as a subhead) that sets italic.

☐ 3. Fractions: Fractions in text are set as piece fractions (½) unless otherwise specified. Fractions in displayed math are set as built-up fractions. Fractions in superscripts are set with a solidus ($x^{1/2}$).

☐ 4. Inferiors and superiors: If a term has both an inferior and a superior, they will be set stacked: P_i^0

☐ 5. Fences: Parentheses, brackets, and braces will be set as tall as is necessary to enclose the entire expression.

☐ 6. In-text equations: All expressions run into general text will be converted to single-line versions, e.g., $\sum_{k=1}^{n}$, $\int_b^a$.

☐ 7. Spacing: We observe the following conventions concerning horizontal spacing in mathematical expressions:

    ☐ No extra space between a number and the symbol it multiplies ($25x$); before or after superscripts, parentheses, braces, brackets, vertical rules, and vertical arrows; and in mathematical expressions in the superior or inferior (x^{1+y}).

    ☐ Thin space or spaceband before and after any symbol used as a verb (such as $= < \sim$); before and after any symbol used as a conjunction ($+ - \times$); before a symbol used as an adjective ($+ -$); after the commas in sets of symbols, sequences of fractions, and coordinates of points ($x_1, x_2, \cdots x_n$); before and after symbols of integration, summation, product, and union; before and after functions (such as cos) set in roman type, except when immediately preceded or followed by fences; before and after vertical rules used singly rather than in pairs; before and after colons used as mathematical symbols and not punctuation; before back inferiors ($a\ _2xb$); and before and after *ds, dp, dx* and similar expressions.

☐ Em space between a symbolic statement and a verbal expression in displayed expressions, and around conjunctions such as "or."

☐ Two-em space between two separate equations or inequalities on the same line of a display, and between a symbolic statement and a condition on that statement.

☐ 8. Breaking equations in text: Here, in order of preference, are the points at which a run-in equation may be broken at the end of a line:

☐ At any 1- or 2-em space (the space will then be deleted).

☐ After a comma or semicolon.

☐ Before or after a symbol used as a verb that does not occur between fences.

☐ Between two sets of fences, whereupon a times sign will be inserted to begin the second line.

☐ For more complex problems, we will follow the guidelines given in *Mathematics into Type* on pages 36–38.

☐ 9. Display equations: Several display equations in succession are considered one display, aligned on the equal sign in each. If there is an intervening word, it is placed at the left margin and the display is considered terminated; any material following the word is then aligned independently. If we have no instructions to the contrary, we will add no extra leading between consecutive equations unless either of them has built-up fractions, radicals, or oversize characters; then we will add a half line space between lines.

☐ 10. Breaking displayed equations: The rules for in-text equations apply in general to displayed equations, except that displayed equations are always broken *before* the operational sign (+, −), which is then aligned to the right of the equal sign. In extremely long equations, where it is necessary to break before the equal sign, the first line of the equation will be flushed left and the second line flushed right. We avoid breaking within fences, but if it becomes necessary, we align the operational sign to the right of the opening fence. For more complex problems, we follow the guidelines given in *Mathematics into Type* on pages 39–41. We will add no extra leading between lines of a runover equation unless either line has built-up fractions, radicals, or oversize characters; then we will add a half line space between lines.

PASTEUP

☐ 11. Avoid a page break between a lead-in line and a displayed equation, even if it means having a widow at the top of the page.

☐ 12. Avoid breaking within a single equation that runs over to 2 or more lines, and never break such an equation to a nonfacing page.

☐ 13. If a page break to a nonfacing page occurs within an Example/Solution or Theorem/Proof, it is better to break within the Example or Solution than between the Example and Solution.

Please list the item numbers which deviate from your desired style and state your preference:

——— ————————————————————————————————

——— ————————————————————————————————

——— ————————————————————————————————

It is probably impossible to give a definition of *table* – exceptions immediately spring to mind, and definitions are not supposed to admit to exceptions. But though *table* cannot be defined, we can point out the sine qua non of tables: they have a vertical axis called the column. Most tables present numerical data and have a structure similar to that of graphs. The many exceptions are usually complicated lists, such as time lines that have column heads but may contain only one horizontal row, or journal entries, where the vertical organization is implied but not formally stated through a column heading. This appendix is concerned with tables that have both horizontal and vertical axes.

Nomenclature

Structurally, tables consist of four general parts: the table number and title, the column heads, the table body, and notes to the table. This structure provides convenient categories for discussing the terms used when designing and typesetting tables, except that the terminology used for table notes is the same as for notes generally and will not be addressed here.

Table Number and Title

Most tables are numbered, allowing them to be placed somewhere near the text reference rather than at the exact point of the callout. Alternatively, a book involving many complex tables, particularly when they are not an integral part of the argument in the text, may have the tables grouped in an appendix. In any case, table numbers are used like footnote numbers, to reference a table to a particular point in the text. The table title, if there is one, gives a brief description of the contents. In writing specifications and marking the manuscript, the terms *table number* and *table title* are usually abbreviated *TN* and *TT.*

Headings over Columns

As noted above, most tables have vertical and horizontal axes, with data placed at the intersections. The vertical components of a table are *columns* (the horizontal ones are *rows*), so the headings over the columns apply to the vertical organization of the table.

There are two basic systems for naming the headings that range over the body of the table. The first of these, by far the more common, identifies the elements by their functions: the term *spanner head* is used for headings that range over more than one column, with *column head* reserved for those that apply to a single column. A spanner head should have a rule below it, a *spanner rule,* that shows just what columns it applies to. The terms *straddle head, straddle rule,* and *crosshead* are sometimes used as synonyms for *spanner head* and *spanner rule.*

The sample below shows the names and functions of these heads:

Table 1. An example of a spanner head and column heads

		SPANNER HEAD	
	Column head, column 1	*Column head, column 2*	*Column head, column 3*
Row one	1.11	2.22	3.33
Row ...	...	...	...
Row *n*	11.11	22.22	33.33

When the terms *spanner head* and *column head* are used in writing specifications, it is understood that the style of the individual heads will be the same for all the tables in a book – if, for example, the column heads are to be set in italic, they will always be italic, whether or not a particular table has spanner heads.

In contrast, the second approach treats column headings like subheads in text – the ordering is not by function, but by perceived importance. Here the heads are usually termed *table column head 1* and *table column head 2*, abbreviated *TCH1* and *TCH2.* Just as an A-head in text may be followed either by text or by another subhead, a first-order head in a table may range over an individual column (if there is only one level of head) or over two or more column heads (if there are two levels). Following are two examples showing the way these heads are typically used, with TCH1 specified to be set in small caps and TCH2 to be set in italic.

A table with two levels of heads:

| | | TCH1 | |
	TCH2, column 1	TCH2, column 2	TCH2, column 3
Row 1	1.11	2.22	3.33
Row ...	...	...	...
Row *n*	11.11	22.22	33.33

compared with one with only one level:

	TCH1, COLUMN 1	TCH1, COLUMN 2	TCH1, COLUMN 3
Row 1	1.11	2.22	3.33
Row ...	...	...	...
Row *n*	11.11	22.22	33.33

Again, naming heads in tables by their function is the more common practice. It has the advantage that a specification sheet giving the typographic style by the function of the heading (e.g., spanner heads to be set in 9-point small caps) removes the necessity of marking the head levels in each table in a manuscript. When the style of the heading is assigned by level of importance rather than function, the head levels in all the tables should be marked.

Occasionally the term *boxheads* is encountered. It refers to a style using vertical rules so that the headings over columns are, in effect, enclosed in boxes. The terms *spanner head* and *column head* are not used; the material a particular heading ranges over is obvious from the box it rests in. With computer-assisted composition the expense of using vertical rules in tables can vary widely, depending on how easy it is to add such rules with a particular typesetting program. Although the style (and the term) is usually considered outdated, an example of a table using boxheads follows:

Table 2. A table with boxheads

	Column head, column 1	Multiple column head	
		Column head, column 2	Column head, column 3
Row one	1.11	2.22	3.33
Row ...	...	...	...
Row n	11.11	22.22	33.33

The Table Body

The table body always contains the data of a table, whether numerical or verbal. As noted above, the term *row* is used to describe the horizontal sequencing; where the leftmost column in some way describes the rest of the data in the row, it is called the *stub*.

Just as the vertical structure of a table may contain several levels of headings (spanner heads and column heads), the horizontal structure may require several levels. Three kinds of heads are used within the table body: sideheads, cut-in heads, and decked heads. Functionally these are usually the same, so that what could be achieved by using one could also be achieved with another. The exception is when a decked head is used not to categorize horizontal information but in effect to change the column headings – usually evidence of a poorly written and edited table.

The term *sidehead* is usually reserved for a head contained entirely within the stub. *Cut-in head* means a head that begins in the stub but, because of its length, is allowed to project into other columns. We should note that some people do not make a distinction between sidehead and cut-in head but use *sidehead* whether or not the head projects beyond the stub. The term *decked head* describes a head that has rules above and below and is frequently centered within the table body rather than beginning in the stub.

Following is a hypothetical table with two levels of heads in the body; the first has to do with ethnic origin and village size, the second with sex. The table is set four ways, showing the various heading styles and, in passing, how one style can be substituted for another.

Using two levels of sideheads:

Table 3   Naming patterns by ethnic group and sex

	Percentage bearing a top-twenty name	Number of names per 100 individuals	Number of unique names
Italians from villages of fewer than 1,000 people			
Females			
Foreign-born Italians	49.3	24.0	15.5
Native-born Italians	47.1	24.8	15.0
Males			
Foreign-born Italians	48.9	19.7	13.5
Native-born Italians	51.6	23.0	14.0
Germans from villages of fewer than 1,000 people			
Females			
Foreign-born Germans	66.6	9.8	4.8
Native-born Germans	52.6	14.7	6.2
Males			
Foreign-born Germans	69.6	8.9	4.3
Native-born Germans	58.8	15.1	6.6

Using a cut-in head and a sidehead:

Table 3   Naming patterns by ethnic group and sex

	Percentage bearing a top-twenty name	Number of names per 100 individuals	Number of unique names
Italians from villages of fewer than 1,000 people			
Females			
Foreign-born Italians	49.3	24.0	15.5
Native-born Italians	47.1	24.8	15.0
Males			
Foreign-born Italians	48.9	19.7	13.5
Native-born Italians	51.6	23.0	14.0
Germans from villages of fewer than 1,000 people			
Females			
Foreign-born Germans	66.6	9.8	4.8
Native-born Germans	52.6	14.7	6.2
Males			
Foreign-born Germans	69.6	8.9	4.3
Native-born Germans	58.8	15.1	6.6

Using a decked head and a sidehead:

Table 3   Naming patterns by ethnic group and sex

	Percentage bearing a top-twenty name	Number of names per 100 individuals	Number of unique names
Italians from villages of fewer than 1,000 people			
Females			
Foreign-born Italians	49.3	24.0	15.5
Native-born Italians	47.1	24.8	15.0
Males			
Foreign-born Italians	48.9	19.7	13.5
Native-born Italians	51.6	23.0	14.0
Germans from villages of fewer than 1,000 people			
Females			
Foreign-born Germans	66.6	9.8	4.8
Native-born Germans	52.6	14.7	6.2
Males			
Foreign-born Germans	69.6	8.9	4.3
Native-born Germans	58.8	15.1	6.6

Using a sidehead and a decked head not projecting into the stub:

Table 3   Naming patterns by ethnic group and sex

	Percentage bearing a top-twenty name	Number of names per 100 individuals	Number of unique names
	Italians from villages of fewer than 1,000 people		
Females			
Foreign-born Italians	49.3	24.0	15.5
Native-born Italians	47.1	24.8	15.0
Males			
Foreign-born Italians	48.9	19.7	13.5
Native-born Italians	51.6	23.0	14.0
	Germans from villages of fewer than 1,000 people		
Females			
Foreign-born Germans	66.6	9.8	4.8
Native-born Germans	52.6	14.7	6.2
Males			
Foreign-born Germans	69.6	8.9	4.3
Native-born Germans	58.8	15.1	6.6

In passing, note that different typefaces, extra space above and below a heading, and indention are alternative ways to show the relative levels of the table body subheads.

In the samples above, all the numbers have been *aligned on the decimal,* whether this decimal is expressed or not. This is a common practice for aligning data within a column in the table body. When rows of data do not relate numerically to each other, however, decimal alignment may not be desirable. Consider the table below (all numbers are fictitious):

Table 4. Esperanto speakers versus English speakers of the world

	Norway	*Brazil*	*Paraguay*
Total population	82,000,000	300,000,000	2,000,000
Percentage speaking Esperanto	0.012	0.06	0
Percentage speaking English	65	23	18

To align the figures expressing percentage on the decimal with those expressing total population would look strange. The rows of a common type – in this case, percentages – do have the data aligned on a common decimal.

Writing Specifications for Tables

It should now be obvious that the number of elements needing specification for tabular composition can be greater than all the other elements in a book. How should these many elements be specified?

A good starting point is for a press to have a house style for tables. Designers may object, with good reason, that a house style will be inappropriate for some books – when, for example, a flush-left style is used for most of the text elements but the house style calls for centered lines with tables or calls for indenting turned lines in the stub, which some feel weakens a flush-left style. Lining figures may be called for in the house style, but for some books old-style figures may be more in keeping with the rest of the text. The obvious solution, it seems to us, is for a press to have a house style with the understanding that a designer may choose not to use it for some titles, at the expense of having to write a complete set of specifications for the tables.

Since tables are graphic, one of the easiest ways to write specifications is to find a complex table that has already been typeset – one at least as complex as any appearing in the manuscript at hand – photocopy the table, and write the appropriate specifications beside each element. Be sure to include the various rules, their weight, and the space above and below the rules.

A second way is to look at every page of the manuscript, identify all the elements that occur in the tables, and write a list that covers them. The first list below is for a book having simple tables – no spanner heads, no heads within the table body, no column heads or stub headings that run more than one line, and no broadside tables. (The abbreviation b/r stands for "base to rule," and the abbreviation r/b stands for "rule to base.")

General: Use old-style figures in number, title, column heads, and notes, lining figures in the body. Normal-aspect tables may be set up to 3 picas wider than text measure.

Table number (TN): 9 pt Clc Sabon bold, flush left.

Table title (TT): 9 pt Clc Sabon roman, run in after TN plus em space.

Rule: ½ pt rule by width of table, 5.5 pts b/r above to TT, 11 pts r/b below to CH.

Column heads (CH): 8/11 Clc Sabon italic, centered in column.

Rule: ½ pt rule by width of table, 5.5 pts b/r above to CH, 11 pts r/b below to first body entry.

Table body: 9/11 Sabon, with italic as marked. Minimum gutter space 1 em; maximum gutter space 4 ems.

Rule: ½ pt rule by width of table, 5.5 pts b/r above, 11 pts r/b below to notes.

Notes: 7/10 Sabon justified by width of table, paragraph style. Paragraph indent 1 em.

Makeup: Tables make up at top of page only, space below to text, 26 pts b/b plus or minus 6.5 pts (may go to 39 pts if necessary). Narrow tables make up flush left on text page; overset tables up to 1 pica into gutter margin and 2 picas into fore-edge margin.

This short list has explicitly covered all the items found in simple tables, including size and placement of the rules. Additionally, implied in the minimum/maximum gutter space, and the specifications for the rules and notes, is the information that narrow tables are not to be set to the width of the text page.

A list of specifications can also be written for more complex tables, having various headings and turned lines. Some manuscript marking on individual tables will obviously be needed, and even with this effort the odds are that something will occur in one or more of the tables that is not covered. In passing, notice that the specifications below employ a strong flush-left orientation.

General: Use old-style figures in number, title, column heads, and notes, lining figures in body.

Table number (TN): 9 pt Clc Sabon roman, flush left, 11 pts below to table title.

Table title (TT): 9/11 Clc Sabon roman, flush left. When title runs more than one line, break so lines are about even.

Rule: ½ pt rule by 26 picas for normal-aspect tables, by width of table for broadside tables, 5.5 pts b/r above to TT, 11 pts r/b below to spanner or column heads.

Spanner heads (SPH): 8/11 Clc Sabon roman, center over columns spanned.

Spanner rule (SPR): ½ pt rule, 5.5 pts b/r to spanner head above, 11 pts r/b below to column head or second spanner head.

Column heads (CH): 8/11 Clc Sabon italic, flush left, turnovers flush left.

Rule: ½ pt rule by 26 picas for normal-aspect tables, by width of table for broadside tables, 5.5 pts b/r above to CH, 11 pts r/b below to table body.

Table body: 9/11 Sabon roman, with italic as marked. Minimum gutter space 1 em.

Sideheads (SH1): 9 pt Sabon roman, flush left, turns flush left, 27.5 pts b/b above, 22 pts b/b below to stub, or 16.5 pts b/b below to SH2.

Sideheads (SH2): 9 pt Sabon italic, flush left, turns flush left, 22 pts b/b above to stub entry, or 16.5 pts b/b above to SH1, 16.5 pts b/b below.

Stubs: Flush left, turns flush left. Extra space above and below as marked.

Rule: ½ pt rule by 26 picas for normal-aspect tables, by table width for broadside tables, 5.5 pts b/r above, 11 pts r/b below to notes.

Notes: 7/10 Sabon justified. All lines flush left. Extra space between notes as marked. Double-column short notes.

Makeup: Tables make up at top of page only, space below to text, 26 pts b/b plus or minus 6.5 pts (may go to 39 pts if necessary).

The second list contains a lot of information – perhaps too much to expect the typesetter and proofreader to follow consistently. The photocopied and marked example recommended above would go a long way toward clarifying the specifications.

Finally, notice that several things have been left out of the "list of specifications." (1) When a stub turns (has two lines), do the data align with the first line of the stub or the last? (2) What should one do if two tables must fall on one page? (3) For tables that take more than one page, how should they break (at a subhead or . . .)? (4) How should the *continued* lines for the table title be worded? Should there be a *continued* line for the subhead if the table must break within the section?

A Few Observations

This is a glossary, not a manual. Nevertheless, just as we have appended some observations to the definitions in the main work, a few observations about preparing tables for composition seem in order.

Many of the design and composition problems that surround tables begin with authors. Unfortunately, many of these problems remain after the editorial work has been done. Editors who do not hesitate to reword an awkward sentence seem reluctant to rework an awkward table. Every editor – indeed, every designer and compositor – would be well advised to read chapter 12 of *The Chicago Manual of Style*. Although it is true that the *Chicago Manual* is the house style for the University of Chicago Press and not Holy Writ, the material covered there must be attended to whether or not one follows Chicago style.

Of particular importance are the section on copyfitting tables (12.75–84) and the section on abbreviations (12.28). One of the most common problems compositors face occurs when a copyeditor marks the abbreviations in column heads to be spelled out. We state here a verity: if the word "percentage" is to appear in six columns of a table, that table will not fit normal aspect on a 26-pica measure, even in 8-point type.

The practice of copyfitting can also alert the editor to other potential production nightmares. When an author has constructed a table with nineteen columns and four rows and just managed to get it to fit broadside on an 8½-by-11-inch manuscript page, there is no possible "design solution" (besides setting the table in 6-point Franklin Gothic Extra Condensed, which most would not consider a *solution*). There is an editorial solution, however: switch the axes of the table so that it has four columns and nineteen rows. Switching axes is covered in section 12.56 of the *Chicago Manual* (illustrated by tables 12.9 and 12.10).

Most editors, designers, and compositors have some prejudices about how material should be presented. These tend to be of two types: protecting the reader, and personal taste about how something should look. From time to time we throw away some of our prejudices, but probably not often enough. An example of the first type is, "if the last word on a recto page is hyphenated, the reader may get lost turning the page." Does anyone seriously believe this? In a like vein, "the *total* line must be indented to separate it graphically from the other stub lines," or "stub lines under a subhead in the table body must be indented, with turns in the stub having a further indention." This may make sense as a matter of style, so that compositors and proofreaders can perform their tasks more easily, but it is hardly a requirement for being able to read and understand a table.

An example of the second type of prejudice is to use lining figures in the body of a table. Lining figures came about in Victorian times, as did the Linotype and Monotype machines. For these machines the lining figures set on a common body, but the old-style figures frequently did not, so lining figures were used in tables simply because they would line up vertically. Most old-style figures in the Linotron 202 typefaces, however, did set on a common body and would line up vertically – in short, there was no longer any manufacturing need to use lining figures in tables. Yet the preference for lining figures in tables remains.

In a different vein, it is possible to design simple tables with no rules at all, by using extra space. If this were done, would it simply be a design affectation, or could it sometimes contribute to the overall look of a book? That's the rub, of course – to distinguish between conventions that preserve good aesthetics and those that are only yesterday's necessary compromises, now somehow elevated to the status of The Way Things Ought to Be.

Table 5. There are no rules*

| | column 1 | SPANNER HEAD | |
		column 2	column 3
Row one	1.11	2.22	3.33
Row ...	...	...	...
Row n	11.11	22.22	33.33

*Many times, however, the best and simplest approach requires rules.

The following list does not pretend to be exhaustive, but it includes many of the accents, diacritics, or special characters commonly encountered in general bookwork. For additional signs and symbols used in mathematics, music, science, and other specialized subjects, consult style guides, scholarly journals, or other authoritative materials in the field.

Á á	a acute
À à	a grave
Â â	a circumflex
Ä ä	a with umlaut/diaeresis
Ã ã	a with tilde
Ă ă	a with breve
Ā ā	a with macron
A̅ a̅	a with longum
Å å	a with overring
Ą ą	a with ogonek/nasal hook/Polish hook
Æ æ	ae ligature/ash/aesc
Ç ç	c cedilla
Č č	c with haček/caron/wedge
Ć ć	c acute
Đ đ	d with crossbar
Ď ď	d with haček/caron/wedge
Ḍ ḍ	d with underdot
É é	e acute
È è	e grave
Ê ê	e circumflex
Ë ë	e with diaeresis
Ĕ ĕ	e with breve
Ẽ ẽ	e with tilde
Ě ě	e with haček/caron/wedge
Ē ē	e with macron
E̅ e̅	e with longum

Ę ę	e with ogonek/nasal hook/Polish hook
Ė ė	e with overdot
ﬀ	ff ligature
ﬃ	ffi ligature
ﬄ	ffl ligature
ﬁ	fi ligature
ﬂ	fl ligature
Ğ ğ	g with breve
Ḥ ḥ	h with underdot
Í í	i acute/faíthe
Î î	i circumflex
Ĭ ĭ	i with breve
Ī ī	i with macron
Ī̄ ī̄	i with longum
Ï ï	i with diaeresis
Ĩ ĩ	i with tilde
ı	dotless i
Į į	i with ogonek/nasal hook/Polish hook
IJ ij	ij ligature
ȷ	dotless j
Ł ł	slashed l (ell)/barred l
Ḷ ḷ	l with underdot
Ṃ ṃ	m with underdot
Ñ ñ	n with tilde
Ň ň	n with haček/caron/wedge
Ń ń	n acute
Ṇ ṇ	n with underdot
Ŋ ŋ	eng
Ó ó	o (oh) acute
Ô ô	o circumflex
Ŏ ŏ	o with breve
Ō ō	o with macron
Ō̄ ō̄	o with longum
Ö ö	o with umlaut/diaeresis

Ø ø	slashed o/barred o
Ő ő	o with tilde
Ǫ ǫ	o with ogonek/nasal hook/Polish hook
Ő ő	o with double acute/Hungarian umlaut
Œ œ	oe ligature/ethel
Ř ř	r with háček/caron/wedge
Ś ś	s acute
Ş ş	s cedilla
Š š	s with háček/caron/wedge
Ṣ ṣ	s with underdot
ß	eszett/German double s
Ţ ţ	t cedilla
Ť ť	t with háček/caron/wedge
Ṭ ṭ	t with underdot
Ù ù	u grave
Û û	u circumflex
Ü ü	u with umlaut/diaeresis
Ũ ũ	u with tilde
Ŭ ŭ	u with breve
Ū ū	u with macron
U̅ u̅	u with longum
Ű ű	u with double acute/Hungarian umlaut
Ů ů	u with overring/kroužek
Ų ų	u with ogonek/nasal hook/Polish hook
Ź ź	z acute
Ž ž	z with háček/caron/wedge
Ż ż	z with overdot
Ð ð	edh/eth
Þ þ	thorn
Ʒ ʒ	yogh
Ƿ ƿ	wyn/wynn/wen
⁊ ⁊	Old English ampersand
Ꝥ	Old English þæt (that)
&	ampersand

*	asterisk
†	dagger
‡	double dagger
§	section mark
‖	parallels
#	number sign/pound symbol/hatchmark
( )	parentheses
[ ]	square brackets
{ }	curly brackets
⟨ ⟩	angle brackets
{ }	brace
¿	inverted question mark
¡	inverted exclamation point
ʔ	glottal stop
¶ ⁋	paragraph mark/pilcrow
« »	guillemets/duck feet/French quotes
ə	schwa
£	pound sterling symbol
¥	yen symbol
◆ ◇	lozenge/diamond
☞	digit/printer's fist/index
∞	infinity symbol

OPERATIONAL SIGNS		TYPOGRAPHICAL SIGNS	
ℐ	Delete	*ital*	Set in italic type
◡	Close up; delete space	*rom*	Set in roman type
ℐ	Delete and close up (use only when deleting letters *within* a word)	*bf*	Set in boldface type
stet	Let it stand	*lc*	Set in lowercase
#	Insert space	*caps*	Set in capital letters
eq #	Make space between words equal; make space between lines equal	*sc*	Set in small capitals
hr #	Insert hair space	*wf*	Wrong font; set in correct type
ls	Letterspace	X	Check type image; remove blemish
¶	Begin new paragraph	V	Insert here *or* make superscript
▯	Indent type one em from left or right	Λ	Insert here *or* make subscript
⌐	Move right		
⌐	Move left	PUNCTUATION MARKS	
][	Center	⌃	Insert comma
⊓	Move up		Insert apostrophe *or* single quotation mark
⊔	Move down		Insert quotation marks
fl	Flush left	⊙	Insert period
fr	Flush right	*set* ?	Insert question mark
=	Straighten type; align horizontally	;	Insert semicolon
‖	Align vertically	*or* :	Insert colon
tr	Transpose	⹀	Insert hyphen
sp	Spell out	M	Insert em dash
		N	Insert en dash
		⌐/⌐ *or* (|)	Insert parentheses

Reprinted from *The Chicago Manual of Style,* 14th edition, © 1969, 1982, 1993 by The University of Chicago.

] Authors As Proofreaders [

["I don't care what kind of type you use for my
]book," said a myopic author to the publisher, but please
print the galley proofs in large type. Perhaps in the
future such a request will not sound so ridiculous]
to those familar with the printing process. today, however, type once set is not reset exepct to correct errors. Proofreading is an Art and a craft. All authors
should know the rudiments thereof, though no proofreader expects them to be masters of it. Watch proofreader expects them to be masters of it. Watch not only
for misspelled or incorrect works (often a most illusive
error, but also for misplace d spaces, "unclosde" quotation marks and parenthesis, and imprper paragraph-
ing; and learn to recognize the difference between an
em dash—used to separate an interjectional part of a
sentence—and an en dash used commonly between
continuing numbers, e.g., pp. 5–10; q.d. 1165 70)
and the word dividing hyphen. Whatever is underlined
in a MS. should, of course, be italicized in print. Two
lines drawn beneath letters or words indicate that these
are to be reset in small capitals three lines indicate
full capitals To find the errors overlooked by the proofreader is the authors first problem in proof reading.
The second problem is to make corrections, using the
marks and symbols, devied by proffesional proofreaders, thay any trained typesetter will understand.
The third—and most difficult, problem for authors
proofreading their own works is to resist the tempta-
tion to rewrite in proofs.

caps + sc   Manuscript editor

1. Type may be reduced in size, or enlarged photographically when a book
is printed by offset.

Reprinted from *The Chicago Manual of Style*, 14th edition, © 1969, 1982, 1993
by The University of Chicago.

Adobe Systems, Incorporated. *Postscript Language Reference Manual.* 2d ed. Reading, Mass.: Addison-Wesley, 1990.

The American Heritage Dictionary of the English Language. 3d ed. Boston: Houghton Mifflin, 1992.

Book Manufacturing Glossary. Ann Arbor, Mich.: Braun-Brumfield, 1991.

Bringhurst, Robert. *Elements of Typographic Style.* Point Roberts, Wash.: Hartley and Marks, 1992.

Butcher, Judith. *Copy-editing: The Cambridge Handbook for Editors, Authors and Publishers.* 3d ed. Cambridge: Cambridge University Press, 1992.

Carroll, John S. *Photographic Lab Handbook.* 5th ed. Garden City, N.Y.: Amphoto, 1979.

CBE Style Manual Committee. *Scientific Style and Format: The CBE Manual for Authors, Editors, and Publishers.* 6th ed. New York: Cambridge University Press, 1994.

Chicago Guide to Preparing Electronic Manuscripts. Chicago: University of Chicago Press, 1987.

The Chicago Manual of Style. 14th ed. Chicago: University of Chicago Press, 1993.

The Compact Edition of the Oxford English Dictionary. 2 vols. Oxford: Oxford University Press, 1971.

Craig, James. *Designing with Type: A Basic Course in Typography.* Edited by Susan E. Meyer. New York: Watson-Guptil, 1971.

———. *Phototypesetting: A Design Manual.* Edited by Margit Malmstrom. New York: Watson-Guptil, 1978.

Dowding, Geoffrey. *Factors in the Choice of Typefaces.* London: Wace, 1957.

———. *Finer Points in the Spacing and Arrangement of Type.* London: Wace, 1954.

Hart, Horace, ed. *Hart's Rules for Compositors and Readers at the University Press, Oxford.* 39th ed. New York: Oxford University Press, 1983.

Hostettler, Rudolf. *Technical Terms of the Printing Industry.* 5th ed. rev. London: Alvin Redman, 1969.

Hurst, C. A., and F. R. Lawrence. *Letterpress: Composition and Machine-Work.* London: Ernest Benn, 1963.

Jacobi, Charles Thomas. *The Printer's Vocabulary.* London: Chiswick Press, 1881.

Jaspert, W. Pincus, W. Turner Berry, and A. F. Johnson. *The Encyclopaedia of Type Faces.* New York: Barnes and Noble, 1970.

Labuz, Ronald. *Typography and Typesetting.* New York: Van Nostrand Reinhold, 1987.

Lawson, Alexander S. *Anatomy of a Typeface.* Boston: Godine, 1990.

———. *Printing Types: An Introduction.* Rev. ed. Boston: Beacon, 1974.

Linotron 202 Font Handling User's Guide. Melville, N.Y.: Mergenthaler Linotype Company, n.d.

McLean, Ruari. *The Thames and Hudson Manual of Typography.* London: Thames and Hudson, 1980.

Mintz, Patricia Barnes. *Dictionary of Graphic Arts Terms.* New York: Van Nostrand Reinhold, 1981.

Pierson, John. *Computer Composition.* New York: Wiley-Interscience, 1972.

Proofreaders' Manual. Baylor, Tex.: G & S Typesetters, n.d.

Pullum, Geoffrey K., and William A. Ladusaw. *Phonetic Symbol Guide.* Chicago: University of Chicago Press, 1986.

Rice, Stanley. *Book Design: Systematic Aspects.* New York: R. R. Bowker, 1978.

Romano, Frank. *Practical Typography from A to Z.* Arlington, Va.: National Composition Association, 1983.

Stone, Bernard. *The Graphic Artist's Illustrated Glossary.* Englewood Cliffs, N.J.: Prentice-Hall, 1987.

Sutton, James, and Alan Bartram. *Typefaces for Books.* Franklin, N.Y.: New Amsterdam Books, 1990.

Swanson, Ellen. *Mathematics into Type: Copy Editing and Proofreading of Mathematics for Editorial Assistants and Authors.* Rev. ed. Providence, R.I.: American Mathematical Society, 1986.

Tschichold, Jan. *The Form of the Book: Essays on the Morality of Good Design.* Point Roberts, Wash.: Hartley and Marks, 1992.

United States Government Printing Office. *Style Manual.* Washington, D.C., 1984.

Updike, Daniel Berkeley. *Printing Types: Their History, Forms, and Use.* 3d ed. 2 vols. Cambridge: Harvard University Press, 1966.

Van Krimpen, Jan. *A Letter to Philip on Certain Problems Connected with the Mechanical Cutting of Punches.* Boston: Godine, 1972.

Webster's New International Dictionary. 2d ed., unabridged. Springfield, Mass.: G. & C. Merriam, 1960.

Webster's Third New International Dictionary of the English Language, Unabridged. Springfield, Mass.: G. & C. Merriam, 1964.

Weiss, Clifford M., ed. *Glossary of Typography, Computerized Typesetting and Electronic Typesetting Terms.* Arlington, Va.: National Composition Association, 1987.

Wilson, Adrian. *The Design of Books.* Salt Lake City: Peregrine Smith, 1974.

Richard Eckersley was educated at Trinity College, Dublin, and the London College of Printing. He joined the University of Nebraska Press as senior designer in 1981 after working at Kilkenny Design Workshops, Ireland. Among the awards he has received are a silver medal from the Leipzig Book Festival and the Carl Herzog Award. He has been visiting lecturer at schools in the United Kingdom and the United States, including the London College of Printing, Maidstone College of Art, Tyler School of Art, and Yale University.

Richard Angstadt started typesetting in 1965, composing books on a Linotype machine and designing theater posters in handset wooden type. Since then he has set up composition systems using ATF phototypesetters, IBM strike-on composers, Mergenthaler VIPs, Linotron 202s, and several versions of laser imagesetters. He is at present kerning Bembo for the seventh time. Since 1985, he has been president of Keystone Typesetting, Inc.

Charles M. Ellertson holds degrees from Saint Olaf College and Duke University. Before becoming a typesetter, he was a recording engineer/producer. While working mainly with music, he also made occasional forays into the worlds of dance and theater. He has been a typesetter for fifteen years and is a cofounder of Tseng Information Systems in Durham, North Carolina. For the past five years he has been experimenting with digital punch cutting, reworking fonts designed for letterpress printing to have good color when printed using the lithographic process.

Richard Hendel is design and production manager at the University of North Carolina Press. He has held the same position at Yale University Press and with other publishers and is also a freelance designer. He has won many awards, including the National Book Award for typographic design and a silver medal from the Leipzig Book Festival. He has taught graphic design at the Rhode Island School of Design, the University of Massachusetts, and the London College of Printing.

Naomi B. Pascal is associate director and editor-in-chief of the University of Washington Press. She was previously an editor at Vanguard Press, New York, and at the University of North Carolina Press. She has written many articles and book reviews dealing with various aspects of publishing and is the author of the article "Copyediting" in the forthcoming *Encyclopedia of Publishing and the Book Arts* (New York: Henry Holt, 1995). In 1991 she received the first Constituency Award from the Association of American University Presses "for her outstanding contribution to scholarly publishing."

Anita Walker Scott is design and production manager at the Johns Hopkins University Press. She was introduced to letters and book design at Cooper Union, New York, and studied calligraphy and type design with Hermann Zapf at AG Stempel in Frankfurt. The recipient of numerous design awards, she has been a freelance designer and a publisher, has been on the staff of major trade publishing houses as well as university presses, and has served as a design and production consultant.